ENVIRONMENTAL PROTECTION IN NEW NUCLEAR POWER PROGRAMMES

The following States are Members of the International Atomic Energy Agency:

AFGHANISTAN	GAMBIA	NORWAY
ALBANIA	GEORGIA	OMAN
ALGERIA	GERMANY	PAKISTAN
ANGOLA	GHANA	PALAU
ANTIGUA AND BARBUDA	GREECE	PANAMA
ARGENTINA	GRENADA	PAPUA NEW GUINEA
ARMENIA	GUATEMALA	PARAGUAY
AUSTRALIA	GUINEA	PERU
AUSTRIA	GUYANA	PHILIPPINES
AZERBAIJAN	HAITI	POLAND
BAHAMAS	HOLY SEE	PORTUGAL
BAHRAIN	HONDURAS	QATAR
BANGLADESH	HUNGARY	REPUBLIC OF MOLDOVA
BARBADOS	ICELAND	ROMANIA
BELARUS	INDIA	RUSSIAN FEDERATION
BELGIUM	INDONESIA	RWANDA
BELIZE	IRAN, ISLAMIC REPUBLIC OF	SAINT KITTS AND NEVIS
BENIN	IRAQ	SAINT LUCIA
BOLIVIA, PLURINATIONAL	IRELAND	SAINT VINCENT AND
STATE OF	ISRAEL	THE GRENADINES
BOSNIA AND HERZEGOVINA	ITALY	SAMOA
BOTSWANA	JAMAICA	SAN MARINO
BRAZIL	JAPAN	SAUDI ARABIA
BRUNEI DARUSSALAM	JORDAN	SENEGAL
BULGARIA	KAZAKHSTAN	SERBIA
BURKINA FASO	KENYA	SEYCHELLES
BURUNDI	KOREA, REPUBLIC OF	SIERRA LEONE
CABO VERDE	KUWAIT	SINGAPORE
CAMBODIA	KYRGYZSTAN	SLOVAKIA
CAMEROON	LAO PEOPLE'S DEMOCRATIC	SLOVENIA
CANADA	REPUBLIC	SOUTH AFRICA
CENTRAL AFRICAN	LATVIA	SPAIN
REPUBLIC	LEBANON	SRI LANKA
CHAD	LESOTHO	SUDAN
CHILE	LIBERIA	SWEDEN
CHINA	LIBYA	SWITZERLAND
COLOMBIA	LIECHTENSTEIN	SYRIAN ARAB REPUBLIC
COMOROS	LITHUANIA	TAJIKISTAN
CONGO	LUXEMBOURG	THAILAND
COSTA RICA	MADAGASCAR	TOGO
CÔTE D'IVOIRE	MALAWI	TONGA
CROATIA	MALAYSIA	TRINIDAD AND TOBAGO
CUBA	MALI	TUNISIA
CYPRUS	MALTA	TÜRKİYE
CZECH REPUBLIC	MARSHALL ISLANDS	TURKMENISTAN
DEMOCRATIC REPUBLIC	MAURITANIA	UGANDA
OF THE CONGO	MAURITIUS	UKRAINE
DENMARK	MEXICO	UNITED ARAB EMIRATES
DJIBOUTI	MONACO	UNITED KINGDOM OF
DOMINICA	MONGOLIA	GREAT BRITAIN AND
DOMINICAN REPUBLIC	MONTENEGRO	NORTHERN IRELAND
ECUADOR	MOROCCO	UNITED REPUBLIC OF TANZANIA
EGYPT	MOZAMBIQUE	UNITED STATES OF AMERICA
EL SALVADOR	MYANMAR	URUGUAY
ERITREA	NAMIBIA	UZBEKISTAN
ESTONIA	NEPAL	VANUATU
ESWATINI	NETHERLANDS	VENEZUELA, BOLIVARIAN
ETHIOPIA	NEW ZEALAND	REPUBLIC OF
FIJI	NICARAGUA	VIET NAM
FINLAND	NIGER	YEMEN
FRANCE	NIGERIA	ZAMBIA
GABON	NORTH MACEDONIA	ZIMBABWE

The Agency's Statute was approved on 23 October 1956 by the Conference on the Statute of the IAEA held at United Nations Headquarters, New York; it entered into force on 29 July 1957. The Headquarters of the Agency are situated in Vienna. Its principal objective is "to accelerate and enlarge the contribution of atomic energy to peace, health and prosperity throughout the world".

IAEA NUCLEAR ENERGY SERIES No. NG-T-3.11 (Rev. 1)

ENVIRONMENTAL PROTECTION IN NEW NUCLEAR POWER PROGRAMMES

INTERNATIONAL ATOMIC ENERGY AGENCY
VIENNA, 2024

COPYRIGHT NOTICE

© IAEA, 2024

Printed by the IAEA in Austria
January 2024
STI/PUB/2076
https://doi.org/10.61092/iaea.vlgp-0prs

IAEA Library Cataloguing in Publication Data

Names: International Atomic Energy Agency.
Title: Environmental protection in new nuclear power programmes / International Atomic Energy Agency.
Description: Vienna : International Atomic Energy Agency, 2024. | Series: IAEA nuclear energy series, ISSN 1995-7807 ; no. NG-T-3.11 (Rev.1) | Includes bibliographical references.
Identifiers: IAEAL 24-01650 | ISBN 978-92-0-155023-1 (paperback : alk. paper) | ISBN 978-92-0-155223-5 (pdf) | ISBN 978-92-0-155123-8 (epub)
Subjects: LCSH: Nuclear power plants — Environmental aspects. | Nuclear power plants — Design and construction. | Environmental protection.
Classification: UDC 621.311.25:502.15 | STI/PUB/2076

FOREWORD

The IAEA's statutory role is to "seek to accelerate and enlarge the contribution of atomic energy to peace, health and prosperity throughout the world". Among other functions, the IAEA is authorized to "foster the exchange of scientific and technical information on peaceful uses of atomic energy". One way this is achieved is through a range of technical publications including the IAEA Nuclear Energy Series.

The IAEA Nuclear Energy Series comprises publications designed to further the use of nuclear technologies in support of sustainable development, to advance nuclear science and technology, catalyse innovation and build capacity to support the existing and expanded use of nuclear power and nuclear science applications. The publications include information covering all policy, technological and management aspects of the definition and implementation of activities involving the peaceful use of nuclear technology. While the guidance provided in IAEA Nuclear Energy Series publications does not constitute Member States' consensus, it has undergone internal peer review and been made available to Member States for comment prior to publication.

The IAEA safety standards establish fundamental principles, requirements and recommendations to ensure nuclear safety and serve as a global reference for protecting people and the environment from harmful effects of ionizing radiation.

When IAEA Nuclear Energy Series publications address safety, it is ensured that the IAEA safety standards are referred to as the current boundary conditions for the application of nuclear technology.

Member States introducing nuclear power programmes face the challenge of building the necessary infrastructure that is essential for the safe, secure, peaceful and sustainable use of nuclear power. IAEA Nuclear Energy Series No. NG-G-3.1, Milestones in the Development of a National Infrastructure for Nuclear Power, first published in 2007, updated in 2015 (Rev. 1), with a second revision forthcoming (Rev. 2), defines three milestones in the development of infrastructure and provides detailed guidance for 19 specific infrastructure issues. The guidance in the publication is referred to as the Milestones approach and is a framework intended to help Member States that are considering or embarking on a new nuclear power programme or expanding an existing one.

Environmental protection is one of the 19 infrastructure issues for which guidance is provided in the Milestones approach. It considers the national infrastructure needed for a new nuclear power programme to ensure the protection of the environment and the studies undertaken to assess, mitigate and manage impacts on the environment. While countries interested in or embarking on new nuclear power programmes are likely to have a national legal and regulatory framework for the protection of the environment, these countries may have little or no experience with the environmental issues specific to nuclear power programmes. To provide information to these Member States, in 2014 the IAEA published IAEA Nuclear Energy Series No. NG-T-3.11, Managing Environmental Impact Assessment for Construction and Operation in New Nuclear Power Programmes. The publication focused on the environmental impact assessment process and the environmental aspects unique to a nuclear power programme.

This publication is a significant revision of IAEA Nuclear Energy Series No. NG-T-3.11. This revision provides detailed information on activities related to environmental protection that are particular to nuclear power, including the legal, regulatory and institutional framework for environmental protection in nuclear power programmes, and the roles and responsibilities of various key organizations. The publication provides an overview of environmental protection activities in each phase of the development of infrastructure for a nuclear power programme and the implementation of environmental management plans during the construction and operation of a nuclear power plant. Although the information on environmental protection provided in this publication is focused on nuclear power plants, it is in general applicable to all nuclear facilities and activities that are included in a nuclear power programme.

The IAEA wishes to acknowledge the valuable assistance of the contributors to this publication, in particular M. Dubinsky (Israel), N. Harman (United Kingdom), D. Herbst (South Africa) and A. Stott (South Africa). The IAEA officers responsible for this publication were J. Haddad and M. Walker of the Division of Nuclear Power.

CONTENTS

1. INTRODUCTION

1.1. BACKGROUND

Energy lies at the heart of both the 2030 Agenda for Sustainable Development and the Paris Agreement on Climate Change. Ensuring access to affordable, reliable, sustainable and modern energy for all will open a new world of opportunities for billions of people through new economic opportunities and jobs, empowered women, children and youth, better education and health, more sustainable, equitable and inclusive communities, and greater protections from, and resilience to, climate change.

The increasing need for energy brings into focus the technology options for low carbon electricity generation. The IAEA, in its rationale and vision for the peaceful uses of nuclear energy, notes that it has the potential to be a reliable, sustainable and environmentally friendly energy source that can contribute to the accessibility of affordable energy services in all interested countries for present and future generations. IAEA Nuclear Energy Series No. NE-BP, Nuclear Energy Basic Principles [1] provides guidance on the basic principle that any use of nuclear energy should be beneficial, responsible and sustainable, with due regard to the protection of people and the environment, non-proliferation and security.

Developing and implementing a nuclear power programme is, however, a major undertaking requiring careful planning and preparation. It requires a major investment in time and in human and financial resources. While nuclear power is not unique in this respect, it is different from other energy sources due to the stringent nuclear safety, security and safeguards requirements associated with the possession and handling of nuclear material and the long term commitment required for a nuclear power programme.

The investment in time and resources is needed not only for the construction of a nuclear power plant (NPP) itself, but also for developing and adopting relevant policies and strategies, establishing appropriate legal and regulatory frameworks, establishing institutions, and developing human resources. In addition, there is a need to enhance existing or establish new physical infrastructure (for example, the national electrical grid, radioactive waste management facilities, environmental monitoring systems), and to undertake a wide range of technical studies and evaluations. All of these activities are included in and known as "development of the infrastructure needed for a nuclear power programme". For a country which does not have nuclear power, it may take up to 10–15 years to develop the necessary nuclear infrastructure.

The IAEA has developed a framework which is intended to help its Member States proceed through the steps necessary to successfully develop a new nuclear power programme. The framework can also assist those Member States interested in expanding an existing nuclear power programme. The framework is described in two key publications — IAEA Nuclear Energy Series No. NG-G-3.1 (Rev. 2), Milestones in the Development of a National Infrastructure for Nuclear Power [2] (hereafter 'the Milestones approach'), and IAEA Nuclear Energy Series No. NG-T-3.2 (Rev. 2), Evaluation of the Status of National Nuclear Infrastructure Development [3]. As explained in Section 2, these publications describe a three phased approach, covering 19 different infrastructure issues that the Member State would need to address, and provide benchmarks ('conditions') against which progress in addressing these issues can be measured.

Environmental protection is one of the 19 infrastructure issues of the Milestones approach [2], and addresses the national infrastructure needed for a nuclear power programme to ensure protection of the environment and studies undertaken to assess, mitigate and manage impacts on the environment.

While countries interested in or embarking on new nuclear power programmes are likely to have a national legal and regulatory framework for the protection of the environment, these countries may have little or no experience of environmental issues specific to nuclear power programmes.

1.2. OBJECTIVE

The objective of this publication is to provide guidance to Member States on environmental protection considerations for new nuclear power programmes. The guidance is also relevant to Member States expanding existing nuclear power programmes. Member States can use this information to ensure that environmental considerations are integrated into the overall decision making process, and that the environmental aspects particular to nuclear power are appropriately addressed during the development of the infrastructure required for a nuclear power programme, and the development and implementation of an NPP project. Guidance and recommendations provided here in relation to identified good practices represent experts' opinions but are not made on the basis of a consensus of all Member States.

1.3. SCOPE

This publication provides detailed information on activities related to environmental protection that are particular to nuclear power, including the legal, regulatory and institutional framework for environmental protection in nuclear power programmes, and the roles and responsibilities of various key organizations. The information provided describes good practices and represents expert opinion but does not constitute recommendations made on the basis of a consensus of Member States.

The publication provides an overview of environmental protection activities in each phase of the development of infrastructure for a nuclear power programme (the Milestones approach [2]) and the implementation of environmental management plans during the construction and operation of an NPP. Although the guidance on environmental protection provided in this publication is focused on NPPs, it is generally applicable to all nuclear facilities and activities that are included in a nuclear power programme.

Several topics related to environmental protection for nuclear power programmes are addressed in more detail in other IAEA publications (the IAEA Safety Standards Series, Safety Reports Series and the IAEA Nuclear Energy Series). Where of relevance, such topics are therefore only briefly discussed in this publication, and reference is made to the appropriate IAEA publication for further information and more specific guidance.

1.4. STRUCTURE

This publication consists of five main sections and two annexes in addition to this introduction. Section 2 provides a high level summary of the Milestones approach [2]. It very briefly introduces the infrastructure issue related to environmental protection and the interconnection with other infrastructure issues of the Milestones approach [2]. Section 3 presents an overview of environmental considerations particular to nuclear power. Section 4 provides an overview of the necessary legislative and regulatory aspects that affect the environmental protection component of the new nuclear power programme. Section 5 describes the environmental protection activities in each phase of the Milestones approach [2], in particular the environmental impact assessment activities and the development of an environmental management plan and an environmental monitoring programme. Section 6 contains the conclusion.

Annex I provides an overview of the expertise typically required for environmental protection processes for nuclear power, while Annex II is a discussion of the management of environmentally related uncertainties that may impact the risks associated with a nuclear power project.

1.5. USERS

This publication is intended for Member States embarking on new nuclear power programmes or expanding existing ones. It will be of primary interest to organizations, managers and officials directly involved in environmental management for new nuclear power programmes, as well as decision makers, advisers, other officials in government, regulatory bodies and technical support organizations (TSOs).

2. THE MILESTONES APPROACH AND ENVIRONMENTAL PROTECTION IN NEW NUCLEAR POWER PROGRAMMES

The Milestones approach [2] has been widely adopted by countries embarking on new nuclear power programmes. The approach defines three milestones in the development of the national infrastructure necessary for introducing nuclear power and identifies specific activities to be conducted in each of the three phases to achieve the respective milestone.

The Milestones approach [2] assumes that the country has evaluated its long term energy needs and supply scenarios and has included nuclear power as a possible supply option in the national energy strategy. The energy planning studies that have been conducted will have taken into account social and economic development goals and the protection of the environment, including climate change mitigation. The development of a nuclear power programme will have been initiated.

A nuclear power programme includes one or more NPPs, radioactive waste processing, storage and disposal, and spent/used fuel storage. It could also include other related activities, such as uranium exploration, mining and processing, and fuel fabrication. The guidance on environmental protection provided in this publication is directly applicable to NPPs, and in general to all nuclear facilities and activities that are included in a nuclear power programme. In general, the scope of environmental impact assessment (EIA) studies and environmental management plans will be similar but may require specific considerations for each type of nuclear facility or related nuclear activity.

The infrastructure needed to support a nuclear power programme encompasses, inter alia, relevant national policies, an appropriate legal and regulatory framework, and an educational and training system to staff key organizations with competent resources. As the programme develops through the three phases of the Milestones approach [2], many specific activities will be undertaken to implement the project for the first NPP.

2.1. THE MILESTONES APPROACH

Figure 1 is a schematic representation of the Milestones approach [2], namely, the phases and milestones for the development of the infrastructure needed for a nuclear power programme. The upper part of the figure represents the development of infrastructure for a nuclear power programme, while the lower part reflects the activities associated with a specific NPP project.

2.1.1. Phases in the Milestones approach

The Milestones approach [2] assumes that a country has already undertaken energy planning studies and consequently has included nuclear power as a possible option to contribute to meeting future energy needs, while addressing national policy and strategic considerations, such as the introduction of low carbon technologies and socioeconomic benefits.

In Phase 1, a series of pre-feasibility and other studies are conducted to enable the country to understand the implications, requirements, international obligations and benefits of a nuclear power

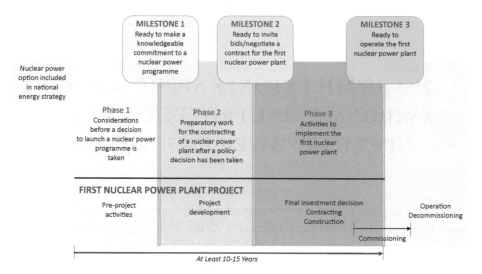

NUCLEAR POWER INFRASTRUCTURE DEVELOPMENT

FIG. 1. Development of the infrastructure for a national nuclear power programme [2].

programme. At the end of Phase 1, a comprehensive report is prepared to enable the government to make a knowledgeable decision on whether to continue and commit to a nuclear power programme.

Phase 2 involves the development of national policies and a legal and regulatory framework for nuclear power, establishment or upgrading of key organizations, development of the required human resources, and completion of key studies. It also involves the confirmation of a site or sites suitable for the proposed NPP and the development of user requirement specifications for engagement with potential suppliers.

Phase 3 is the period during which the financial and contractual commitments are made and the construction of the NPP is carried out. Section 2.2 and Section 5 provide further guidance on environmental protection activities in each of the three phases.

2.1.2. Key organizations

The Milestones approach [2] identifies three key organizations as the main players in the development of the nuclear power programme infrastructure:

(1)   The government and its nuclear energy programme implementing organization (NEPIO), established early in Phase 1.

The NEPIO is a mechanism to coordinate the work of all organizations involved in the development of nuclear infrastructure. The role of the NEPIO is described in IAEA Nuclear Energy Series No. NG-T-3.6 (Rev. 1), Responsibilities and Functions of a Nuclear Energy Programme Implementing Organization [4]. The government ministry or department responsible for environmental matters is typically represented on the NEPIO.

(2)   The regulatory body or bodies responsible for nuclear safety, nuclear security, safeguards, radiological environmental impact, non-radiological environmental impact and protection, and non-radiological occupational health and safety.

Appropriate agreements are typically implemented between the regulatory bodies to ensure the coordination and consistency of their respective regulatory functions and activities, in particular with respect to the review and authorization of the EIA in Phase 2, the environmental management plan and monitoring programme for construction in Phase 3, and thereafter for operation and eventual decommissioning of the NPP.

(3)   The owner/operator of the proposed NPP.

From the perspective of environmental protection, the responsibilities of the owner/operator in Phase 2 include conducting the EIA in compliance with the country's environmental laws, regulations and the international legal instruments to which the country is a State Party. The owner/operator is responsible to ensure that the bid invitation or user requirements specification document for engagement with potential suppliers contains environmental information and requirements, in particular, any legally binding requirement. The owner/operator is also responsible for ensuring the implementation of all requirements specified in the environmental management plan. Further guidance on the role and responsibilities of the owner-operator is provided in IAEA Nuclear Energy Series No. NG-T-3.1 (Rev. 1), Initiating Nuclear Power Programmes: Responsibilities and Capabilities of Owners and Operators [5].

Each of the key organizational entities has a specific role to play, with responsibilities changing as the programme advances. Typically, consultants and/or TSOs will be appointed by the key organizations to assist them in the execution of their respective responsibilities, particularly early in the development of the infrastructure when sufficient expertise may not be available in the organization. Section 4.3 further describes organizational aspects related to environmental protection in the development of the infrastructure for a nuclear power programme.

2.1.3. Infrastructure issues

The Milestones approach [2] defines a set of 19 infrastructure issues to be addressed during each of the three phases of the development of the infrastructure required for a nuclear power programme (see Fig. 2). The 19 infrastructure issues include both "hard" infrastructure such as the national electrical grid and the site where the proposed NPP could be constructed, and "soft" infrastructure such as national policies, laws and regulations, and human resources development.

Publications listed in the Nuclear Infrastructure Bibliography [6] describe and provide guidance on activities associated with the 19 infrastructure issues.

FIG. 2. The 19 infrastructure issues of the Milestones approach [2].

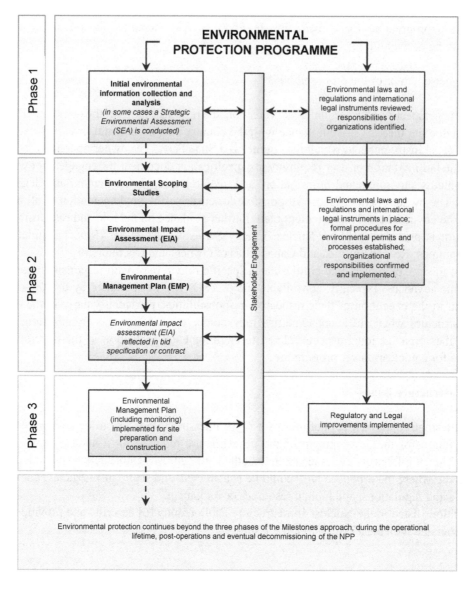

FIG. 3. Steps of the environmental protection programme within the Milestones approach [2].

2.2. ENVIRONMENTAL PROTECTION WITHIN THE MILESTONES APPROACH

Environmental protection is one of the 19 infrastructure issues of the Milestones approach [2]. Figure 3 shows the key steps of a phased implementation of an environmental protection programme within the framework of the Milestones approach [2].

2.2.1. Phase 1

In Phase 1, a country considers environmental aspects that will influence the decision to proceed (or not) with a nuclear power programme. Some countries may require that a strategic environmental assessment (SEA) is conducted for the nuclear power programme. The SEA consolidates and assesses relevant environmental aspects in a structured approach, and generally involves engagement with stakeholders including the public. Guidance on conducting an SEA for a nuclear power programme is provided in IAEA Nuclear Energy Series No. NG-T-3.17, Strategic Environmental Assessment for Nuclear Power Programmes: Guidelines [7].

In Phase 1, the NEPIO will also review the country's existing national legal and regulatory framework for environmental protection and its international obligations and, if necessary, develop plans for amendments to ensure the suitability for a nuclear power programme. Steps related to environmental laws and regulations are discussed in more detail in Section 4.

A survey of the country is undertaken to identify regions and potential sites for an NPP and select candidate sites for further study. The environmental attributes and criteria [8, 9][1] for use in the site survey and site selection processes are established. The collection and analysis of the available initial environmental information is performed either as a component of the site survey and selection process or through an SEA.

The comprehensive report prepared by the NEPIO at the end of Phase 1 reflects the outcome of an SEA (if conducted) or the results of an analysis of the initial environmental information gathered, and the approved environmental attributes, exclusion and avoidance criteria to be used in selecting the most appropriate site. The comprehensive report also includes a summary of the review and recommendations on possible enhancements or clarifications in existing environmental laws, regulations and responsibilities.

2.2.2. Phase 2

Early in Phase 2, environmental information collected from field studies, or an SEA (if conducted), supports the identification of the preferred site(s). The outcome of an SEA (if conducted) also provides a framework and information for the EIA process.

The EIA process is conducted in accordance with national requirements for the identified site(s). An environmental management plan (EMP) incorporating mitigation measures identified during the EIA process is prepared. The EMP forms both the guideline and a legally binding document addressing environmental issues throughout the project life cycle from pre-construction to operation, and further aims to address and respond to uncertainties identified during the EIA. Further guidance on the EIA process and the EMP is provided in Section 5.

The environmental authorization or record of decision (if available), the EIA report (including specification of the site environmental conditions, characteristics and data), and the EMP are used in preparation of the bid invitation specification or contract.

Ideally, in this phase, decision making and licensing processes are confirmed or established. Responsibilities between the nuclear regulatory body, environmental regulatory body and other key organizations are confirmed or allocated.

The enhancements or clarifications needed in existing environmental laws, regulations and organizational responsibilities that were identified during the review in Phase 1 (see Section 2.2.1) to ensure their suitability for a nuclear power programme, and recommended in the comprehensive report, are implemented. During this phase, the authorizations required for construction are identified.

2.2.3. Phase 3

Early in this phase, the various authorizations for environmental requirements are obtained by the project proponent and additional activities as necessitated by authorizations for construction and operation are conducted. During this phase, the EMP for construction is implemented. The collection of baseline data continues and the environmental monitoring programme for construction is implemented as part of the EMP to monitor the impacts during construction and confirm the impacts are as expected or, if not, take remedial actions. The EMP for operation is reviewed and updated, if necessary.

---

[1] Attributes are issues, events, phenomena, hazards and other aspects to be considered in the relevant studies and evaluations. Attributes are further described in IAEA Nuclear Energy Series No. NG-T-3.7 (Rev. 1), Managing Siting Activities for Nuclear Power Plants [8], and IAEA Safety Standards Series No. SSG-35, Site Survey and Site Selection for Nuclear Installations [9].

During this phase, audits, and, depending on the requirements, possibly external independent audits, are carried out on a regular basis on behalf of the authorities and lenders to confirm compliance with the conditions of the environmental authorization as documented in the EMP.

2.2.4. After Phase 3

The environmental protection programme continues during the operational lifetime of the NPP; in particular, the EMP for operation is implemented. Environmental protection would also be required and implemented for eventual decommissioning of the NPP. Further information is provided in Sections 3.4, 4.3.1, 5.2.2.2 (d), and 5.3.

2.3. INTERACTION WITH OTHER INFRASTRUCTURE ISSUES

The infrastructure issue "Environmental protection" does not stand alone, since there is an interconnection with several other infrastructure issues identified in the Milestones approach [2]. The following is a brief description of the interconnection of environmental protection with other relevant infrastructure issues.

2.3.1. National position

The national position is the outcome of a process that establishes the governmental strategy and commitment to develop, implement and maintain a safe, secure and sustainable nuclear power programme. IAEA Nuclear Energy Series No. NG-T-3.14, Building a National Position for a New Nuclear Power Programme [10] provides guidance in this regard. At the end of Phase 1, a comprehensive report is prepared to enable the government to make a knowledgeable decision and commitment to the nuclear power programme, if that is the recommendation. The comprehensive report is based on the results of pre-feasibility and other relevant studies and technical evaluations, and addresses both potential risks and benefits, for example, in the mitigation of climate change. It includes initial environmental information and approved environmental attributes and criteria used in relevant studies and evaluations, as well as a summary of the review and recommendations on possible enhancements or clarifications in existing environmental laws, regulations and responsibilities. If available, a summary of the results of an SEA at the programme level is incorporated into the comprehensive report.

2.3.2. Nuclear safety

There are areas of overlap between the EIA for the proposed NPP and the safety analysis, in particular with regard to NPP parameters and the site characterization and analysis. Baseline physical information about the site and its environs are required by both the EIA and the site licensing processes. Several specialist studies required for site characterization for the site evaluation report are also needed for the EIA, for example, geotechnical, hydrological, seismic and meteorological studies. A prospective radiological impact assessment will be conducted in Phase 2 in preparation for the site authorization application to the nuclear regulatory body. Depending on national requirements, a summary of the results or the full report of this radiological impact assessment could be incorporated into the EIA report.

Depending on the national EIA regulations, studies and data required for the safety analysis report for evaluation and approval by the nuclear regulatory body, for example population demographics, dispersion of radioactive material and contamination, the range of quantities and the management of radioactive waste, and a high level assessment of potential nuclear accidents would be summarized in the EIA report. The assessment of the probability and impact of natural external hazards, such as seismic or severe meteorological events, and the potential impact of climate change on the safety and sustainability of the

NPP would form part of both the site licensing and the EIA processes. The agreement between the nuclear and environmental regulatory bodies would determine how such topics are evaluated and approved.

Care should thus be taken to ensure consistency of data, information and conclusions between the EIA, site characterization and nuclear safety assessment processes.

2.3.3. Management

Effective management of a nuclear power programme from study through implementation to operation of the NPP is crucial to its success. The implementation of appropriate management systems in all relevant organizations, including quality assurance and control systems, is an integral element of nuclear programmes, and a requirement of nuclear regulatory bodies. The environmental authority may require organizations whose activities impact the environment to implement a management system and, in some cases, to obtain and maintain formal certification, for example the International Organization for Standardization Environmental Management System, ISO 14001:2015 [11].

In executing its activities related to nuclear infrastructure development, it is appropriate for each of the key organizations to implement a management system that includes environmental data and decisions. The owner/operator will also ensure that the bid invitation or user requirements specification document, for engagement with potential suppliers, contains environmental information and requirements, in particular any legally binding requirement.

2.3.4. Funding and financing

As part of the development of the required nuclear infrastructure, it is necessary that sufficient funds are available in Phase 1 to gather and analyse initial environmental information or conduct an SEA (if required), and in Phase 2 for conducting the EIA and associated specialist studies. Such funds do not form part of the financing of the NPP itself and would be provided by the organization responsible for the EIA, normally the owner/operator as part of the feasibility studies for the project.

A thorough analysis of environmental protection requirements, including required mitigation solutions, helps to reduce the uncertainty associated with the financial risks. By including the environmental protection requirements and mitigation solutions in the bid invitation or user requirements specification document, the possibility of unplanned and costly environmental protection measures impacting the design is minimized.

As a condition for a financing arrangement, many financial institutions/lenders require that appropriate environmental policies and a management system are in place, evidence of meeting environmental performance criteria is available and an EIA has been conducted and approved by the competent authority.

Funding mechanisms for the management of radioactive waste and spent fuel and for future decommissioning of the plant and remediation of the site are also expected to be in place by Milestone 3.

2.3.5. Legal and regulatory framework

It is expected that most countries embarking on a new nuclear power programme will have a legal and regulatory framework for environmental protection in place, and that any nuclear facility developed as part of the programme will need to comply with all existing, applicable environmental laws and regulations. However, certain laws or regulations may need to be supplemented or amended, or new laws and regulations may be required to fully address the environmental issues resulting from a nuclear power programme.

Roles and responsibilities at national and local level are expected to be clear, and appropriate agreements implemented between the regulatory bodies to ensure the coordination and consistency of their respective regulatory functions and activities, for example, in respect to the review and authorization of the EIA for the proposed NPP in Phase 2. Section 4 provides further description of the environmental legal and regulatory framework and international legal instruments applicable to nuclear power programmes.

2.3.6. Radiation protection

Radiation protection pertains to the protection of workers and the public during the construction and operation of the NPP. A description of the radiation protection programme is included in the safety analysis report submitted to the nuclear regulatory body in support of the application for a nuclear authorization. The nuclear regulatory body would also provide oversight to ensure the implementation of the radiation protection programme, and in particular, compliance with the national radiation dose limits for workers and the public. Depending on the national environmental regulations, an EIA report prepared for a proposed NPP may contain a section that briefly describes the national radiation dose limits, the radiation protection programme and the oversight role of the nuclear regulatory body. The EMP is likely to include a requirement to implement the radiation protection programme and a radiation surveillance and monitoring programme for the environment.

2.3.7. Electrical grid

Depending on the national environmental regulations, consideration of the environmental impact due to upgrade or expansion of the electrical power grid is likely to be included in a separate EIA, which may be the responsibility of an organization other than the owner/operator of the proposed NPP. Although the EIA for the proposed NPP and the EIA for any upgrade or expansion of the electrical grid may be carried out separately, it is preferable that they are conducted in parallel and that potential cumulative environmental impacts are considered.

Electrical supply during construction of the NPP will require line connections and substations. The environmental impact of these connections and substations would typically be included in the EIA for the proposed NPP.

While selecting sites for the NPP, scoping of the environmental impacts of upgrades or expansion of the electrical power grid (line corridors and substation sites) associated with the NPP will ensure that there are no environmental fatal flaws related to the grid that could eliminate a site at a later stage.

2.3.8. Human resource development

The knowledge and competencies needed by the NEPIO, owner/operator, environmental regulatory body and environmental practitioners, respectively, related specifically to the environmental impact and requirements of nuclear power, can initially, if not nationally available, be obtained through appropriate contracts with international consultants and TSOs. However, in such situations, the organization is expected to have sufficient knowledge to specify the work to be outsourced as well as to understand the results presented by the subcontractors in order to be a "knowledgeable customer". International outsourcing contracts typically include provisions for capacity building and the transfer of knowledge to local staff to ensure the sustainability of the nuclear power programme. Annex I describes the types of expertise required to develop and implement an environmental protection programme for nuclear power.

2.3.9. Stakeholder engagement

A proposal to develop a new nuclear power programme will inevitably result in debate, nationally and locally, when specific sites have been identified, and potentially also with neighbouring countries. Consequently, countries embarking on new nuclear power programmes should expect to spend significantly more resources and time on engaging with stakeholders on environmental protection matters than those that may be expected for or associated with other industries, activities or projects.

A broad range of stakeholders are involved in the development of a new nuclear power programme, including governments and communities in neighbouring countries. Some countries require that an SEA or its equivalent, which inherently includes stakeholder participation, is conducted to support decision making at the energy policy level or at a programme level, for example, for the introduction of nuclear power. In most countries, the legal and regulatory framework will include provisions for stakeholder engagement during SEA and EIA processes. The regulations may include a requirement that an organization, independent of the owner/operator for the proposed NPP, coordinates and manages public participation in the SEA and/or EIA processes.

If consultations with neighbouring and other States are conducted, the timelines for the SEA and/or EIA process and stakeholder engagement activities would take into account the differences between the national legal and regulatory systems and timelines for completion of consultations with the other States involved. The time required for document translation may also significantly affect the overall schedule. In planning the stakeholder engagement process, the competent authority would consider these complex issues.

Each of the key organizations involved in the nuclear power programme should develop and implement stakeholder engagement plans, and, to the extent consistent with security and commercial requirements, practise transparent and open communication regarding the programme. Further guidance on stakeholder engagement and the development and implementation of stakeholder engagement strategies and plans are provided in IAEA Nuclear Energy Series No. NG-G-5.1, Stakeholder Engagement in Nuclear Programmes [12]. Specific guidance on stakeholder engagement and public participation during an SEA is included in IAEA Nuclear Energy Series No. NG-T-3.17, Strategic Environmental Assessment for Nuclear Power Programmes: Guidelines [7]. Guidance specifically for nuclear regulatory bodies, related to informing stakeholders of their activities while maintaining their independence of nuclear energy promotional activities, is provided in IAEA Safety Standards Series No. GSG-6, Communication and Consultation with Interested Parties by the Regulatory Body [13].

2.3.10. Site and supporting facilities

There is a clear connection between environmental protection and site selection, which necessitates the coordination of data collection and the conduct and use of studies that are required by both the siting and the EIA processes. The site survey and selection process for NPPs is strongly influenced by specific environmental considerations for the regions and sites of interest. The site selection process narrows down the site options according to established environmental, safety related and other criteria. Consideration of environmental issues early in the site selection process would result in a selected site that is acceptable with regards to the environmental and socioeconomic impacts. In addition to the impact of the construction and operation of the proposed NPP on the environment, the likelihood and impact of natural external hazards, such as seismic or severe meteorological events, and the potential impact of climate change on the safety and sustainability of the NPP, as well as other factors, are considered in the site selection process. IAEA Nuclear Energy Series No. NG-T-3.7 (Rev. 1), Managing Siting Activities for Nuclear Power Plants [8] provides more guidance and SSG-35 [9] gives recommendations on the siting process.

More detailed information and data are collected and detailed analyses are conducted for evaluation of the selected site(s). Safety related requirements for site evaluation are established in IAEA Safety Standards Series No. SSR-1, Site Evaluation for Nuclear Installations [14]. Similarly, detailed information and data are collected and detailed analyses are conducted for the EIA for the proposed NPP. Several studies conducted for the characterization of the selected site are closely related to studies required for the EIA.

In addition to the proposed NPP itself, there may be additional supporting facilities constructed at the site. These can include spent nuclear fuel and radioactive waste storage facilities, electrical switchyards and transformer stations, and new access roads or barge facilities. The impacts associated with the construction and operation of such supporting facilities on or in the immediate vicinity of the site should be included in the EIA for the proposed NPP (upgrades or expansion to the electrical power grid are discussed in Section 2.3.7).

2.3.11. Emergency planning

NPP safety systems are designed to minimize the probability of a large release of radioactive material from the plant. The probability is not zero, however, and previous accidents have demonstrated that emergency planning for the protection of plant personnel, emergency workers and the public beyond the site boundary is a necessary element of overall plant safety.

Emergency planning ensures the capability to take actions that will effectively mitigate the consequences of an emergency. Environmental data and conditions, including but not limited to geological and topographical data, typical meteorological data and atmospheric dispersion potential, land use, population densities and distributions, transport and communication infrastructure, and the probability and impact of natural external hazards such as extreme meteorological events, provide input into the development of the nuclear emergency preparedness and response (EPR) plan submitted for approval to the nuclear regulatory body.

Depending on national requirements and the coordination agreements between the nuclear and environmental regulatory bodies, the EIA report for the proposed NPP may a summary of the legal and regulatory framework for EPR for nuclear accidents, the concept of an emergency plan to mitigate the accidental release of radioactive material and aspects such as the technical basis and typical emergency planning zones.

2.3.12. Nuclear fuel cycle

The nuclear fuel cycle for an NPP requires uranium mining and processing, conversion and enrichment, fuel element fabrication and spent fuel storage and disposal. In some cases, the reprocessing of spent fuel is also included in the nuclear fuel cycle. Transportation of nuclear material will be required. Most of these aspects are likely to be located outside countries embarking on new nuclear power programmes. The national environmental legal and regulatory framework would provide guidance on the level of detail required to address this in the overall EIA process.

2.3.13. Radioactive waste management

The production, interim storage on-site, transport to and storage or disposal at off-site facilities of non-radioactive waste will be addressed in the EIA for the proposed NPP in accordance with the country's requirements and environmental regulations. The production, interim storage on-site, transport to and storage or disposal at off-site facilities of radioactive waste and spent fuel is subject to authorization by the nuclear regulatory body. Depending on national requirements and the coordination agreements between the nuclear and environmental regulatory bodies, a summary of radioactive waste management may be included in the EIA report for the proposed NPP.

2.3.14. Industrial involvement

Many commodities, components and services are required to construct and support the operation of an NPP. Such supporting activities can be a source of jobs and economic growth for the country. The opportunity for industrial involvement in the nuclear power programme is likely to have been included in an SEA (if conducted). Depending on the national environmental regulations, the opportunity for national and local industrial involvement can be addressed in the socioeconomic studies that form part of the EIA for the proposed NPP.

2.3.15. Procurement

It will be necessary to procure the services of an experienced, competent organization to conduct an SEA (if required) and the EIA for the proposed NPP. It is also likely to be necessary to procure equipment to perform the baseline data gathering and ongoing monitoring of specified environmental parameters. National procurement legislation and regulations will establish the requirements and provide guidelines and may require amendment to meet the needs of the nuclear power programme.

3. ENVIRONMENTAL CONSIDERATIONS PARTICULAR TO NUCLEAR POWER

In many respects, environmental considerations for nuclear power are quite similar to those for other industries and large projects. However, there are some aspects of environmental protection that are particular to nuclear power and that are not normally considered for other projects. These include: (a) direct exposure to sources of radiation in the NPP; (b) exposure to radionuclides released to the environment during normal operation; and (c) the potential for accidents with concomitant release of radionuclides to the environment and the need to put in place EPR plans around the NPP. The management of radioactive waste and spent nuclear fuel from NPP operations are also specific to nuclear power. These particularities are described in this section.

3.1. RADIATION AND RADIOACTIVITY RELEASED INTO THE ENVIRONMENT

Exposure to radiation is the most obvious aspect of nuclear power that is different from most other power supply options.[2] Exposure can occur externally as a result of radiation from unshielded radioactive materials used in the NPP, or due to radionuclides released to the environment and deposited onto soils and surfaces. Exposure can also be internal, from radionuclides that have been discharged to the environment from the NPP and taken into the body through inhalation, ingestion or dermal absorption. External and internal exposures could be due to a planned operational release or the result of an accidental release. There are data and analysis techniques available to assess the impacts from such exposures. Information on established methods and criteria for estimating and assessing impacts on humans, flora and fauna from these types of exposures is provided in IAEA Safety Standards Series No. GSG-10, Prospective Radiological Environmental Impact Assessment for Facilities and Activities [15], IAEA Safety Standards Series No. NS-G-3.2, Dispersion of Radioactive Material in Air and Water and Consideration of Population Distribution in Site Evaluation for Nuclear Power Plants [16], IAEA Safety Standards Series No. SSG-79, Hazards Associated with Human Induced External Events in Site Evaluation for Nuclear Installations [17] and Safety Reports Series No. 19, Generic Models for Use in Assessing the Impact of Discharges of Radioactive Substances to the Environment [18].

3.1.1. Direct exposure to sources of radiation during normal operation

During normal operation of NPPs, direct radiation exposure is important only if the exposed individuals are very close to the highly radioactive materials. To avoid such situations, nuclear facilities are designed so that direct radiation from the radioactive material is blocked by the plant's steel and

---

[2] Some non-nuclear industries can release naturally occurring radioactive materials (NORM) to the environment; for example, during the extraction and use of fossil fuels such as coal, oil and natural gas.

concrete structures (containment) and other materials. Access to areas in the facility where there would be exposure to direct radiation is prevented by physical measures and administrative procedures. Activities that may result in direct exposure to radiation are carefully analysed and planned, and protection measures are implemented, including the training of workers to minimize any exposure. Workers on the site and visitors to the site are monitored and appropriate measures are taken to limit their exposure to below the applicable constraints. Direct radiation exposure to individuals off-site is negligibly small.

As mentioned in Section 2.3.6, occupational radiation exposure is addressed in the infrastructure issue on radiation protection. Depending on the national environmental regulations, an EIA report prepared for a proposed NPP may contain a section that briefly describes the national radiation dose limits, the radiation protection programme and the oversight role of the nuclear regulatory body. The EMP is likely to include a requirement to implement the radiation protection programme and a radiation surveillance and monitoring programme.

3.1.2. Exposure to radionuclides released to the environment during normal operation

Under controlled conditions, NPPs release very small amounts of radioactive gases and liquids into the environment. Radioactivity can also leave the NPP site in the form of solid waste shipped off the site. These releases of radionuclides to the environment are subject to a process of authorization by the regulatory body and monitored to ensure that they pose no danger to the public or the environment.

Releases to the atmosphere occur through stacks or vents on top of buildings and to surface water bodies through release points and design provisions approved by the nuclear regulatory body. These releases are calculated on the basis of dissipation into the environment with infinite dilution options such as a large water body and where the level of natural background radiation normally remains unaltered as a result of releases from the NPP. Nevertheless, the quantity and composition of radioactivity released during normal operation is an important input for estimating the radiological environmental impacts from the NPP when compared to the environmental monitoring data gathered before commencement of operation. Other inputs required for such studies are the typical environmental data required for transport modelling, such as topography, hydrological and meteorological characteristics, and data regarding, inter alia, population distribution and habits, and uses of land and water in the region of the site. The natural background radioactivity at and around the site is measured and recorded prior to operation, so that the potential incremental differences of the radioactivity in the environment due to the NPP can be measured. A methodology for assessing the radiological impacts on the public and the environment during normal operations for facilities and activities is described in GSG-10 [15], with elaboration for NPPs provided in NS-G-3.2 [16].

3.1.2.1. Meteorology (atmospheric dispersion modelling and dose assessment)

Meteorological studies require the longest time frames to acquire data. These studies are important in assessing both the impact of the environment on the plant and the impact of the plant on the environment. For the former, the frequency of rare or extreme meteorological events and how they may affect the safety and operation of the plant are determined; these may also involve assessment of climate change over the lifetime of the project. For the latter, both routine and accidental releases need to be considered, and meteorological conditions representative of the NPP site are determined in terms of wind speed and direction, atmospheric stability and precipitation, all of which can affect the dispersion of routine or accidental radioactive releases to the environment and their subsequent impact on people and the environment. This will also be an input to emergency planning and transboundary assessments.

In Phase 1, the site survey is undertaken, during which regional climate data are collected and analysed to determine potentially suitable sites or regions, for which the meteorological extremes and the likelihood of rare meteorological events are likely to be acceptable. Candidate sites are identified (see IAEA Nuclear Energy Series No. NG-T-3.7 (Rev. 1), Managing Siting Activities for Nuclear Power Plants [8] and SSG-35 [9]). During the EIA in Phase 2, site meteorological data will be considered in the air quality and dispersion modelling studies. Experts in climatology will also need to consider how the meteorological extremes and the likelihood of rare meteorological events will change over the lifetime of the NPP project (~100 years) as a result of climate change. More detailed information on meteorological studies, the modelling and analysis of the dispersion of radioactivity in the atmosphere can be found in NS-G-3.2 [16] and in IAEA Safety Standards Series No. SSG-18, Meteorological and Hydrological Hazards in Site Evaluation for Nuclear Installations [19].

3.1.2.2. Surface and groundwater transportation modelling and dose assessments

Liquid released to the environment during the normal operation of NPPs is usually discharged to a surface water body in the vicinity of the NPP. This body could be a river, a lake of natural or anthropogenic origin, or the marine environment. Similar to the atmospheric transport calculations, an environmental transport analysis is conducted for the radionuclides in surface water from the point of release to points accessible to human and environmental receptors. If the water is used for irrigation, uptake by plants and transfer from the plants to humans and from the plants to animals and subsequently to humans (through the food chain) is accounted for. Uptake of radionuclides by aquatic biota and transfer to humans and terrestrial animals is also considered.

Although there could be contamination of groundwater with small quantities of radionuclides during normal operation, there are no planned releases to groundwater. Under conditions of normal operation, only extremely small quantities of radionuclides that are released to air or surface water bodies may reach groundwater through deposition from the air onto ground, followed by infiltration through soil with precipitation or through the interaction of surface water bodies with the groundwater.

More detailed information and guidance on the modelling and analysis of the transport of radioactivity in surface water and groundwater can be found in NS-G-3.2 [16].

3.1.3. Radiological impacts of accidents

Accidents can happen in any industry or large project. NPPs have an additional consideration to take into account, the potential for release of large quantities of radioactivity during some of those accidents. Accidents can be caused by natural and human induced external events, such as earthquakes, tornadoes, tsunamis, airplane crashes, or by internal events, such as equipment failure and human error. NPPs incorporate a series of engineered safety features into their designs to minimize both the probability of occurrence of accidents and the quantities of radionuclides released to the environment in the event of an accident. Through a set of extensive analyses, the possible accidents, their probabilities of occurrence, the quantities, physical and chemical form of radionuclides released to the environment, the release point and the time profile of the release are determined and studied. The dispersion and radiation exposure analyses are similar to those conducted for normal operations, except that the environmental data (e.g. wind speed and direction and stability class) are for short term releases and more conservative in their assumptions, and the computational models are more complex than the ones used for normal operation. These types of analyses are conducted primarily during Phase 2 and SSG-79 [17] provides guidance on assessing external human induced events when conducting a site evaluation for an NPP.

More detailed information on the modelling and analysis of the dispersion of radioactivity in the atmosphere, surface water and groundwater can be found in NS-G-3.2 [16].

3.2. EMERGENCY PLANNING

Before the start of operation, the operators of the NPP — in cooperation with local and national authorities — will need to develop a plan for an emergency response in the event of an accidental release of radioactivity to the environment. Such a plan will detail how various countermeasures, such as evacuation, sheltering, decontamination of individuals, iodine thyroid blocking, restrictions on the consumption of food, milk and drinking water, control of access and traffic restrictions will be implemented. Environmental data will be key inputs in developing the emergency plan, for example local residential, worker, and temporary populations, transport and communication infrastructure, and local food production. Special population groups who would require additional measures to be put in place to perform countermeasures such as evacuation — for example, hospitals, schools, care homes and prisons — would also need to be considered. Such measures are established in IAEA Safety Standards Series No. GSR Part 7, Preparedness and Response for a Nuclear or Radiological Emergency [20].

As mentioned in Section 2.3.11, this topic is addressed in the infrastructure issue on emergency planning. Environmental data and conditions provide input into the development of the nuclear EPR plan submitted for approval to the nuclear regulatory body or other competent authority. Depending on national requirements and the coordination agreements between the nuclear and environmental regulatory bodies, a summary of the legal and regulatory framework for EPR for nuclear accidents, the concept of an emergency plan to mitigate accidental release of radioactive material and aspects such as the technical basis and typical emergency planning zones may be included in the EIA report for the proposed NPP.

3.3. MANAGEMENT OF RADIOACTIVE WASTE

The management of radioactive waste and spent nuclear fuel, at the site or the transport of these materials away from the site, will need to be considered. An overview of the possible technical options for managing radioactive waste is given in IAEA Nuclear Energy Series No. NW-G-1.1, Policies and Strategies for Radioactive Waste Management [21].

The management of liquid and gaseous radioactive wastes is of particular importance, given that the management processes could result in effluents containing radioactive material, and which requires an authorization for discharges to the environment.

In most countries, a separate SEA and/or EIA would be required for off-site facilities for the processing, storage or disposal of radioactive waste, and the storage and disposal of spent nuclear fuel. The need for such facilities and the potential impacts that could be associated with them would be acknowledged in the EIA report for the NPP. If studies have already been completed that provide detailed analysis of the environmental impacts for existing or proposed off-site facilities in the country, the studies would be referenced and the results summarized in the EIA report for the proposed NPP.

3.4. DECOMMISSIONING

An SEA, if conducted at the nuclear power programme level, would consider possible decommissioning options, including radioactive waste management strategies and ensuring the provision of funding for eventual decommissioning. The EIA for a specific project will need to demonstrate that the NPP can be decommissioned with an acceptable environmental impact. Since decommissioning is likely to commence many decades into the future, after 60–80 years of operation, it is not addressed in detail in the EIA report for construction and operation of an NPP, but it is described in principle, with the currently available possibilities for decommissioning.

In most countries, a separate specific EIA would be required for the decommissioning of the NPP, which would consider, amongst others, the processing and/or disposal of radioactive waste, the physical rehabilitation or repurposing of the site, and monitoring of the situation until the release of the site from regulatory control. Rehabilitation to clean up contamination on a site is a major aspect of decommissioning which requires detailed investigation and management, with removal or containment of contamination as may be appropriate. Decommissioning policies and strategies that involve radioactive materials are examined in more detail in IAEA Nuclear Energy Series No. NW-G-2.1, Policies and Strategies for the Decommissioning of Nuclear and Radiological Facilities [22].

4. LEGAL, REGULATORY AND INSTITUTIONAL FRAMEWORK FOR ENVIRONMENTAL PROTECTION IN NUCLEAR POWER PROGRAMMES

Countries embarking on a new nuclear power programme will need to ensure that their legal and regulatory framework appropriately accounts for the radiological and non-radiological environmental aspects of the programme, along with the alignment and integration of these two distinct but equally important areas.

The legal framework includes but is not limited to the national constitution and laws, which are influenced by international guidelines and practices contained in various conventions. It establishes the environmental responsibilities of the various organizations involved in and necessary for a successful nuclear power programme. The regulatory framework includes environmental regulations, policies, decisions, guidelines, procedures and authorizations which are developed to support the constitution and laws of the country. In addition, financial institutions have environmental requirements which they prescribe or provide as guidelines when funding projects.

4.1. INTERNATIONAL LEGAL INSTRUMENTS

Several international and regional legal instruments (i.e. conventions, treaties, and agreements) exist and these collectively focus on general and specific environmental principles and decision making practices, the prevention of environmentally harmful practices, including climate change, pollution limitation and remediation, and the preservation of natural resources and biodiversity. Although these instruments are legally binding only on the States Parties (i.e. those countries that have acceded to them), consideration of the ethics and principles defined by the relevant instruments are also useful for non-party States in their development of a national nuclear power programme.

The development of general environmental principles and decision making was initiated in 1972 with the United National Declaration Conference on the Human Environment (Stockholm Declaration) and continued with the 1992 United Nations Conference on Environment and Development in Rio de Janeiro. The latter resulted in the Rio Declaration on Environment and Development, setting the ground for the main methodological principles of environmental protection by specifying implementation means for environmental protection. The Rio Declaration defines public participation as a way to better handle environmental issues and develops the idea that environmental damages should be compensated for, as well as the precautionary principle that the "lack of full scientific certainty shall not be used as a reason for postponing cost effective measures to prevent environmental degradation". The Rio Declaration also stresses that the execution of EIAs is necessary before activities or projects likely to have a significant adverse impact on the environment are undertaken.

Countries that are States Parties to the Convention on Access to Information, Public Participation in Decision Making and Access to Justice in Environmental Matters (Aarhus Convention, 1998) [23] are required by the convention to make provision for a number of rights of the public, including access to environmental information and justice, the right to participate in the development of public policy related to the environment and the right to participate in environmental decision making. This aligns to the requirement to undertake public participation during EIAs and to ensure it is meaningful.

The Convention on Environmental Impact Assessment in a Transboundary Context (Espoo Convention, 1991) [24] requires its States Parties to provide information and afford authorities and the public in affected neighbouring countries the opportunity to participate in its (the States Parties') EIA.

Beyond giving general methodological guidelines, a number of international legal instruments were developed to protect particular areas or components of the environment or to prevent certain activities or behaviour which may damage the environment. These legal instruments encompass two different goals: the prevention of environmentally harmful practices, including through pollution limitation and remediation, and the preservation of resources. The following are examples of international legal instruments for the prevention of pollution and the preservation of nature and the ecosystem (the year of adoption and it's is specified in brackets):

— Convention on Wetlands of International Importance especially as Waterfowl Habitat (Ramsar Convention, 1971) [25];
— Convention on the Prevention of Marine Pollution by Dumping of Wastes and Other Matter (London Convention, 1972) [26];
— Montreal Protocol on Substances That Deplete the Ozone Layer (Montreal Protocol, 1987) [27];
— Basel Convention on the Control of Transboundary Movements of Hazardous Wastes and their Disposal (Basel Convention, 1989) [28];
— Bamako Convention on the Ban of the Import into Africa and the Control of Transboundary Movement and Management of Hazardous Wastes within Africa (Bamako Convention, 1991) [29];
— Convention for the Protection of the Marine Environment of the North-East Atlantic (OSPAR Convention, 1992) [30];
— United Nations Framework Convention on Climate Change (1992) [31];
— Convention for the Protection of the Marine Environment and the Coastal Region of the Mediterranean (Barcelona Convention, 1995) [32];
— The Paris Agreement under the United Nations Framework Convention on Climate Change (Paris Agreement, 2015) [33].

Many of the most important environmental impacts of an NPP are non-radiological in nature. For countries that are States Parties to conventions, treaties or regional agreements on environmental protection, the non-radiological aspects are often dealt with under such international legal instruments and, as a norm, are integrated into national legislation. Although these instruments are obligatory only to the States Parties, it might be useful for countries that are not party to consider the ethics and principles defined by the relevant instruments for possible positive use in the development of a national nuclear power programme and incorporation into the country's legislation, if applicable.

The relevant international binding legal instruments concerning radiological protection of people and the environment under the auspices of the IAEA include:

— Convention on Early Notification of a Nuclear Accident (1986) [34];
— Convention on Assistance in the Case of a Nuclear Accident or Radiological Emergency (1986) [35];
— Convention on Nuclear Safety (1994) [36];
— Joint Convention on the Safety of Spent Fuel Management and on the Safety of Radioactive Waste Management (1997) [37].

To the extent that they address compensation for nuclear damage to the environment in case of accidents, nuclear liability conventions can also be considered significant in the context of environmental protection. Examples include:

— Paris Convention on Third Party Liability in the Field of Nuclear Energy (1960) [38];
— Vienna Convention on Civil Liability for Nuclear Damage (1963) [39];
— Convention on Supplementary Compensation for Nuclear Damage (1997) [40].

The IAEA safety standards provide the fundamental principles, requirements, recommendations and guidance to ensure nuclear safety. They serve as a global reference for protecting people and the environment and contribute to a harmonized high level of nuclear safety worldwide. The IAEA safety standards establish the fundamental safety objective and principles of protection and safety, and, through safety requirements, impose conditions that should be met to ensure the protection of people and the environment, both now and in the future. Through safety guides, they also provide recommendations and guidance on how to comply with the requirements.

4.2. FINANCIAL INSTITUTION AND LENDER REQUIREMENTS

Large projects most often require funding sourced from various financial institutions or lenders. These lenders have environmental, social and safety requirements with which the proponent of the project is expected to comply in order to receive the necessary funding for the project. The requirements often include aspects related to biodiversity, water, air quality, land management, climate change, resettlement, impact on indigenous people and cultural heritage. Further, the requirements include the full range of phases of a project, from assessment and monitoring to pre-construction, construction and commissioning to operations and eventual decommissioning and site rehabilitation.

The specific requirements that are applied depend on the country's own regulations and how these compare with the lender's requirements. Generally, the lenders would opt for stricter standards. There are several environmental, social and safety guidelines from lenders that could be applicable through the phases of a project, for example, those of the World Bank, the International Finance Corporation, or the Organization for Economic Cooperation and Development (OECD). Financial institutions in more than 30 countries have also adopted what are known as the Equator Principles which are "a financial industry benchmark for determining, assessing and managing environmental and social risk" [41].

4.3. LEGAL AND REGULATORY INTEGRATION FOR A NUCLEAR PROGRAMME

It is expected that most embarking countries will have a framework for environmental protection in place, and that any nuclear facility developed as part of the programme will need to comply with all existing, applicable environmental laws. A sound legal framework provides developers and investors with a clear understanding of expectations, thus helping to reduce project risks.

In certain cases, there may be a need for additional environmental laws and regulations to be developed or supplemented to achieve the appropriate level of environmental laws and regulations and to facilitate the integration with nuclear safety requirements. Additionally, newcomer countries will need to ensure that organizations responsible for implementing environmental laws and nuclear safety have been determined, are adequately resourced and have roles and responsibilities that are clearly defined. If the country is not yet competent on issues related to the environmental aspects of nuclear power, they could seek support from countries experienced in implementing a nuclear power programme.

When augmenting environmental laws, countries consider the harmonization of nuclear and non-nuclear related environmental laws, and ambiguities between the different laws and the regulators are identified and clarified. In this manner, conflicting or duplicate legal requirements will be resolved at an early stage before they impede the programme development.

While many countries may already have comprehensive legislation requiring environmental protection and assessment of environmental impacts for major projects, other laws and regulations may also be required to fully ensure the protection of the environment. For example, separate laws may address the protection of human health, water (groundwater, surface fresh water and sea water), land and soil, air and the atmosphere, and biodiversity and ecosystems. Additionally, laws on cultural heritage, environmental justice and socioeconomic issues are important aspects of the legal framework for environmental protection inclusive of social justice. Therefore, laws addressing the full range of environmental impacts are expected to be in place — or under development — before a knowledgeable decision is made to proceed with a nuclear power programme.

Depending on the national law, the nuclear and environmental components for site selection and licensing of the proposed NPP site could be carried out by the same regulatory body. However, the environmental components are often the responsibility of a separate environmental authority or regulatory body. In the event that the regulatory bodies are separate, it is advisable to implement a legally binding cooperative agreement between the environmental and nuclear regulatory bodies. Such an agreement would set out the roles and responsibilities of the regulatory bodies, and how these bodies would integrate the various licensing processes.

Giving consideration to the dual but distinct programmes of nuclear safety and environmental authorization, which cannot always be executed in parallel, it may be required by the environmental regulatory body to include some elements of radiological studies into the EIA. In terms of radiological impacts, the focus of the EIA process is to assess the potential impacts of radiological releases (including normal operational releases and accident conditions).

4.3.1. Application of environmental protection laws throughout the programme

The applicability of environmental laws and regulations to the project and its influence on decision making will increase in depth and detail as the project progresses through the phases of the Milestones approach [2].

During Phase 1, a thorough review and suitability of the existing legal framework for environmental protection, as well as policy and planning, including aspects such as local and international policies, plans, guidelines, cooperative agreements, international and regional conventions and treaties, is undertaken. The results of this review and recommendations to close identified gaps would be included in the comprehensive report for the nuclear power programme.

During Phase 2, the policy and legal requirements form an important component of the EIA and specialist studies and would influence the final approval to proceed with the NPP at a particular site. Generally, the legislation prescribes the process to be followed and dictates the scope and depth of detail to be adhered to, in order to obtain the required environmental authorization from the relevant authority to proceed with the project. During Phase 2, the EMP is developed by the project proponent, which documents the mitigation measures to be implemented during the construction and operation of the plant. The EMP will describe the role of all stakeholders and identify the various regulators applicable in the next phase of the project.

The start of NPP construction in Phase 3 takes the project to the next level of detail, most often prescribed in regulations, guidelines and procedures. Apart from the environmental and nuclear authorizations at the start of construction, any other authorizations would be obtained during Phase 3. This can generally involve several different regulatory bodies, depending on the country's governance structure.

After Milestone 3 has been achieved, startup and the associated final commissioning would commence, leading to operation of the NPP. Environmental protection activities would continue during the operational phase, in particular the mitigation measures and monitoring specified in the EMP for operation.

4.3.2. Licence, authorizations and permits

The legal framework in a country generally includes different forms of licences, authorizations and permits, a number of which would be applicable for an NPP. For the purposes of this publication, the terms 'licence', 'authorization' and 'permit' in the context of environmental governance are considered to be synonymous and in this publication will be referred to as 'authorizations.'

Examples of other environmental authorizations which may be required, in addition to the approval of the EIA and the nuclear authorization, are:

— Waste management authorization (transport, storage and disposal of non-radioactive waste);
— Waste disposal site authorization (for general waste if the disposal site is not a government-licensed site);
— Water use authorization (including disturbance of water resources);

— Sewage work authorization;
— Coastal management discharge authorization;
— Fresh water discharge authorization.

Conditions of operation as outlined in the relevant 'authorization' may vary depending on the country's legislative framework, but it is of great importance to formulate the conditions precisely and unambiguously. Measurable parameters and respective quantifications should be used as much as possible to define whether actions comply with the rules and authorization. Conditions can be based on discharge limits, but they can also be based on outcomes. Clear conditions help avoid disputes, problems and unforeseen costs at later stages of construction, commissioning and operation of the NPP.

Attention would also be paid to issues related to the validity of the 'authorization', confirmation of compliance and corrective measures, such as:

— Renewal periods;
— Monitoring programmes;
— Review of findings and recording;
— Reporting requirements;
— Appropriate actions when conditions are breached;
— Inspection authority, frequency and costs;
— Enforcement mechanisms.

Depending on how developed a site is, additional authorizations may be required for construction of related infrastructure projects, such as roads, transmission lines or waterways. Many of the authorizations may be interconnected and dependent on each other (e.g. approval of one application for authorization is given provided that another is already in force), so a delay in obtaining one authorization may have a domino effect and result in serious delays in the overall NPP project schedule (of up to a few years). Additionally, approval of the EIA report and the entire EIA process by the competent authority may well be necessary before the nuclear authorization is awarded. Therefore, sufficient resources for the regulatory process for the variety of required authorizations need to be allocated. Relevant government departments or competent authorities would be consulted as early as possible in the development of the nuclear power programme.

4.4. ORGANIZATIONAL ASPECTS

The requirements for a clear governance framework and adequate organizational capacity are critical for a successful nuclear power programme and are expected to be developed early, taking into account the key role players and stakeholder engagement, the integration of radiological and environmental aspects, and the project schedule requirements.

As described in Section 2, the Milestones approach [2] identifies three organizational entities which are regarded as key for a successful nuclear programme: (a) the government and the NEPIO that it establishes; (b) the regulatory authorities, including those for environmental protection; and (c) the owner/operator of the future NPP. Each step of the nuclear power programme requires input or decision making by one or more of these organizations. The NEPIO plays a key role in terms of governance of the nuclear power programme through the establishment of a steering committee, which would include representation from regulatory bodies including those mandated with environmental protection.

Since environmental governance is an integral and critical component of the development of the infrastructure for a nuclear power programme, it is essential to ensure that all stakeholders understand and execute their role in developing the environmental protection process, integrating this with energy planning, siting and nuclear safety, carrying out sound and comprehensive impact assessments, and making timely decisions.

Environmental protection is achieved through guidance and control performed by numerous government organizations and executed by the owner/operator. Although one organization typically has overall responsibility for environmental protection and the EIA decision making process, other government departments or organizations could be mandated through the national legal and regulatory framework for related matters covered in the EIA specialist studies, such as the protection of water resources, agriculture, tourism and socioeconomics. In such cases, it is important that formal agreements, such as memorandums of understanding, are developed between the organizational entities, which define the working arrangements, responsibilities and accountabilities. In particular, a memorandum of understanding is required to confirm in a prescriptive and detailed manner the respective roles and responsibilities of the nuclear regulatory body and the environmental regulatory body.

In Phase 1 of the Milestones approach [2], the focus lies on assessing the capability of national organizations to deal with issues such as the preparation of guidelines, responsibilities and capacities for completing and reviewing aspects relevant to environmental protection, the necessary authorization steps and interfaces, and coordination of environmental activities. The NEPIO, the nuclear regulatory body — if it exists — and the environmental regulatory body are usually involved in this step, and an action plan to resolve identified deficiencies is developed. The action plan would identify who would develop and implement the legal and regulatory requirements, taking into account the necessary independence, as well as procedural relations among the suggested responsible organizations. These are important elements in minimizing potential conflicts of interest that may affect the programme, including energy planning, radiological safety and environmental protection. Prior to initiating Phase 2, which includes the EIA, it is important that all relevant and required guidance and related documentation is in place. Capacity building, if required, would need to be initiated and advanced sufficiently to allow for further development during Phase 2.

In Phase 2, the relevant regulatory bodies would continue to develop regulations and guidelines, which are applied during Phase 2 and Phase 3 and operation. During this phase, the governance processes are expected to be in place to evaluate and authorize the EIA. Towards the end of Phase 2, the regulatory bodies need to ensure that environmental authorization processes are clearly defined and that there is organizational capacity to assess and approve environmental authorizations. The EIA is conducted during Phase 2, usually by the owner/operator if the necessary skills are available in-house or by an independent consultant engaged by the owner/operator.

In Phase 3, the regulatory bodies would continue to develop regulations and guidelines applicable during construction and operation of the plant. The regulatory bodies would also acquire adequate skills for compliance monitoring during construction. The owner/operator would ensure the availability of appropriate skilled resources, internal and external, for the execution of the EMP.

Strong and independent governance and a clear understanding of the roles and responsibilities of all organizations, in particular regarding compliance monitoring, is required throughout the construction, operation and decommissioning of the NPP.

5. NUCLEAR ENVIRONMENTAL PROTECTION — PROCESS AND ACTIVITIES

As mentioned in Section 4.3, most countries considering a nuclear power programme will have an existing environmental protection framework in place, identifying the processes and activities for projects, from their initiation through to their implementation, and eventually the termination of activities and post-activity responsibilities. The general steps of the environmental protection process in each phase of development of a nuclear power programme, as discussed in the Milestones approach [2], can be found in Section 2.2. Specific expertise is required for environmental protection activities in general, and specifically for nuclear power. Annex I provides an overview of the expertise needed for environmental protection processes for nuclear power.

In general, the environmental protection activities in Phase 1 will typically take about 9–12 months, but could take longer, depending on the governmental decision process. Phase 2 activities may take several years to properly scope the study effort, collect the data, conduct the EIA process including stakeholder engagement, and obtain all reviews and approvals. The schedule is affected by many variables, including specific national requirements, the time needed for stakeholders to provide input on their environmental concerns and for the proponent to develop appropriate methodologies to address them, and the time to collect and evaluate the necessary data (i.e. through all seasons and life stages of the flora and fauna).

Environmental management and monitoring are implemented in Phase 3 for the site preparation, NPP construction and initial commissioning. These environmental management and monitoring activities continue throughout the entire period of plant operations as well as during decommissioning, a period that extends over several decades. The environmental protection processes and activities in each phase are described in greater detail in this section.

5.1. ENVIRONMENTAL PROTECTION ACTIVITIES IN PHASE 1

5.1.1. Initial environmental information

As mentioned in Section 2.2, in Phase 1, the country (usually the NEPIO) would gather initial environmental information and establish environmental attributes and criteria that would be used in a survey to identify regions of interest and potential sites for NPPs (see IAEA Nuclear Energy Series No. NG-T-3.7 (Rev. 1), Managing Siting Activities for Nuclear Power Plants [8] and SSG-35 [9]). This is mainly carried out through a desktop study of available information. Environmental information previously used in energy planning studies could form the starting point for the gathering of this initial environmental information. As the number of sites being scrutinized against exclusionary and discretionary criteria is reduced, the amount and detail of environmental information collected for each remaining site may increase. The quality and completeness of the data used in these studies are not generally subject to regulatory review, and while sufficient for these early purposes, the data will need to be verified by the collection of additional data using strict quality control measures during the EIA. Data gaps identified in this phase will be filled later in the EIA process (during Phase 2).

The initial environmental information process extends from Phase 1, with the data gathered to evaluate the suitability of potential sites, to the early stages of Phase 2, when the study intensifies with the collection of additional data and interpretation, leading to the selection of a preferred site or sites for the EIA studies.

All available archived data on land use, historical and cultural resources, meteorology and air quality, geology, hydrology, ecology, socioeconomics and environmental justice, the radiological and chemical environment, and related national projects will be considered in the initial environmental information report. In most cases, however, there are significant gaps in the data on the site itself, even if general information is available. These gaps are documented and will be filled later in the process and certainly before the EIA is considered complete. A study of the presence of archaeological or cultural artefacts is often undertaken by specialists early in the overall programme, as some Member States do not permit any development in areas where such artefacts are present.

As the initial environmental information analysis takes place in Phase 1, its development would typically be the responsibility of the NEPIO. In some cases, depending on the approach and/or requirements of the individual country, a separate nuclear power programme SEA (see IAEA Nuclear Energy Series No. NG-T-3.17, Strategic Environmental Assessment for Nuclear Power Programmes: Guidelines [7]) may be completed in Phase 1. The SEA is a structured process to identify environmental concerns and perform a high level consideration of the environmental issues likely to be significant to a particular project. It brings specific attention to environmental concerns early in the strategic or programmatic development of a project. Depending on the Member State requirements, the SEA may vary in scope and the time required to complete it. The results of an SEA, if conducted, would be summarized in the comprehensive report at the end of Phase 1, and would provide input into the scoping report in the EIA process in Phase 2.

As mentioned in Section 2.2.1, a summary of the initial environmental information and approved environmental attributes and criteria used in relevant studies and evaluations would be included in the comprehensive report compiled by the NEPIO at the end of Phase 1.

5.1.2. Older environmental information or impact assessments

In some countries, initial environmental information may have been gathered, a survey conducted and potential sites identified several years earlier. A country that already has an existing nuclear power programme could be interested in expanding a site to accommodate one or more new plants co-located with existing NPPs. In these cases, earlier work (and an earlier EIA, if conducted) is considered part of the baseline information and reconsidered for the current project to take into account changed environmental conditions, including socioeconomic conditions, changed regulations and requirements, and new technologies governing data collection and evaluation.

5.2. ENVIRONMENTAL PROTECTION ACTIVITIES IN PHASE 2

The environmental authorization for a proposed project is typically the outcome of the application process originated by the project proponent and submitted to the relevant authority, supported by documentation on the assessment and mitigation of impacts on the environment of the project and environmental management and monitoring plans.

5.2.1. Environmental scoping studies

In Phase 2, the environmental scoping process is initiated once a decision has been taken to proceed with the nuclear power programme. The environmental scoping process is the process of identifying key environmental issues across alternative sites. The scoping process identifies all aspects and likely significant environmental issues for which there may be impacts to be assessed as well as the requirements for additional information and analysis in order to complete a comprehensive EIA of the planned project. It builds on the information provided in the initial environmental information as the starting point for planning the scope of the necessary analyses to understand site specific environmental issues and, as appropriate or necessary, evaluate site alternatives for the proposed NPP project.

During the planning phase, all available data sources are identified (perhaps confirming what has already been completed earlier, such as for an initial environmental information report or feasibility study, if one has been prepared), relevant issues are determined and taken into consideration, study areas are defined and general methodologies for additional data collection and analysis are specified. A thorough walk-through or inspection of the preferred site(s) and environs will likely be necessary to determine whether there are potential issues to be considered, and which have not become apparent from the initial desktop studies and available information.

The report prepared as a result of this process is frequently termed the environmental scoping report (ESR). It includes available NPP technical information, specifies how the environmental investigations and impact assessments are to be conducted (including methodologies), identifies issues of special interests, describes legal and regulatory interfaces, and identifies stakeholders both initially consulted and those to be consulted in the future. It would also define whether the EIA report is to include an assessment of additional items, such as the nuclear fuel cycle front end and/or back end, transmission lines and roads. Additionally, identified alternatives as well as the consequences of not constructing the project (the so-called 'no go option') need to be agreed for further discussion in the EIA. Thus, the ESR provides a roadmap for conducting the EIAs which will eventually result in the preparation of an EIA report.

An important purpose of the scoping process is to identify and communicate with key stakeholders (as well as regulatory authorities) and gather local knowledge related to potential environmental impacts. It is recommended that the stakeholders be contacted early in the process of developing the ESR and their concerns taken into consideration, so that the scoping study includes all the items deemed important or relevant. The ESR thus provides important input to a stakeholder participation and communication plan. After the initial draft scoping report is prepared for stakeholder review, including review by the public if this is part of the national process, consultations are held with the competent authority and perhaps some other agencies. Once the comments have been incorporated into a revised draft, the public may be involved to gather further comments. Input from stakeholders at any stage may necessitate revising the scope of the environmental studies to address their concerns. After all the comments have been received and considered, the report is finalized and approved by the competent authority.

5.2.1.1. *Typical contents of the environmental scoping report*

The format of an ESR may vary depending on national requirements or preferences. Since it will be reviewed by decision makers and stakeholders, it should contain information about the site, the project, and the environmental conditions and issues as well as a justification of the project with respect to the country's energy plan.

The report typically includes an overall discussion of the environmental protection and EIA processes and the relationship between the scoping report and the EIA, including the timelines and the scope of work, a list of regulatory required authorizations throughout the steps, and the intended public participation process and stakeholders to be involved.

The description of the project contains as much detail as possible based on the information available at that time. Both construction and operational aspects would be described, particularly interfaces with the environment (impacts on the environment and impacts of the environment on the project). In many cases, the nuclear power technology and vendor, size or location on the site selected for the proposed NPP will not have been decided at the time of development of the ESR, and perhaps not even at the time of preparation of the EIA report. In this case, the plant parameter envelope[3] (PPE) principle, in which values estimated to have greater negative impact or lesser positive impact on the environment than those that might be realized in the selected technology, without being overly conservative, can be used for the environmental assessment. These bounding values are selected based on the best information about the potential technologies and sites being considered. When the exact technology and site information are known, the actual values of the parameters necessary for the EIA are compared with the PPE values used. If the actual values exceed the bounding PPE values, the impact on the relevant parts of the EIA will be analysed to determine whether a reassessment using the actual values is needed.

An important consideration to be estimated during this time is the potential environmental impact area of the project, sometimes referred to as the region of influence, based on the available information. The impact area will be different for different impact categories (e.g. for socioeconomics, cultural resources, ecological impacts) as well as for releases to air, surface water, and groundwater, and would be estimated with a considerable margin to result in the area(s) of concern to be included within the study area. In the absence of known data regarding the expected technology, the PPE as described above can be used as source emission terms and simple models as described in Section 5.2.2, along with a qualitative inspection of the surrounding terrain, and can be used to define a potential impact area. It is possible that different sites may require consideration of different impact areas due to differences in the

---

[3] Plant parameter envelope: a tabular set of data that addresses all technologies under consideration and identifies each aspect leading to a potential environmental impact, along with the value associated with each technology. The PPE includes the important physical and chemical parameters that may affect the environment (such as water requirements, land use and emissions) for the considered plants, and identifies the highest impact value or range of values for each parameter. These bounding parameters are then used for environmental analysis in the EIA process.

presence/absence of surface water bodies or groundwater, or terrain which alters dispersal of airborne contaminants. Additionally, the impact area may vary with the type of technology.

National requirements often require that alternative options to the planned NPP project are described in the EIA. These alternatives could include alternative sites for the construction of the NPP, other energy alternatives (e.g. fossil fuels, renewable energy, importation from another country or region), alternative plant and transmission systems (e.g. once-through cooling versus cooling towers, different intake or discharge locations on the cooling reservoir).

The available baseline environmental information is included in as much detail as possible, including a description of the study area for each preferred site considered. The baseline information can be presented in maps, figures and tables. Data gaps (which vary widely depending on whether the site has been studied before and upon the type of resource being evaluated) would be identified in order to enable the design of a data collection programme.

The process for baseline environmental data collection, including the methodology to gather the required information to fill data gaps, would be provided in enough detail to ensure that sufficient data, of appropriate quality, is collected. This includes a description of sampling locations and frequency, as well as the general time schedule for data collection, since some types of data can be collected even before the scoping report receives final approval, while others may require extended time frames for collection.

A baseline environmental data collection programme includes all the elements required for a comprehensive industrial EIA report but is more extensive. For an NPP project, the potential for radiological impact is a continuing concern. Therefore, specific studies are performed to understand the baseline occurrence of radiological contamination, including from sources naturally occurring in the environment as well as from sources that are the result of human activity (including releases from weapons testing activities in the past and other NPPs around the world). It is important to establish the baseline occurrence in the absence of the plant under consideration in order to determine later the incremental contribution of the proposed NPP on the radioactivity in the environment, as some of these radiological components may be very similar. The time period over which data are collected should be sufficiently long to allow variations and trends to be identified – in other words, the expected evolution of the baseline in the absence of the NPP project needs to be determined. Baseline monitoring programmes will vary depending on the receiving environment and the nature and extent of potential impact. IAEA Safety Standards Series No. RS-G-1.8, Environmental and Source Monitoring for Purposes of Radiation Protection [42], provides guidance on environmental and source monitoring for the purposes of radiation protection, and further information is provided in Safety Reports Series No. 64, Programmes and Systems for Source and Environmental Radiation Monitoring [43]. More detailed information on the modelling and analysis of the dispersion of radioactivity in the atmosphere, surface water and groundwater can be found in NS-G-3.2 [16].

The methodologies used for the baseline environmental data collection are established on a case by case basis. They define how to assess the significance of the impacts, how the data will be used and how uncertainties are handled, including in the technology design.

The public participation plan is typically described in the ESR, including the identification of stakeholders and discussion of the process for stakeholder engagement. Guidance on stakeholder engagement is provided in IAEA Nuclear Energy Series No. NG-G-5.1, Stakeholder Engagement in Nuclear Programmes [12].

5.2.2. Environmental impact assessment

In Phase 2, the EIA for the proposed NPP is performed by or under the responsibility of the project proponent. The principles of EIA best practice discussed in the Principles of Environmental Impact Assessment Best Practice of the International Association for Impact Assessment (1999) [44] define an EIA as "the process of identifying, predicting, evaluating and mitigating the biophysical, social, and other relevant effects of development proposals prior to major decisions being taken and commitments made".

The objectives of an EIA are (see Ref. [44]):

— To ensure that environmental considerations are explicitly addressed and incorporated into the development decision making process;
— To anticipate and avoid, minimize or offset the adverse significant biophysical, social and other relevant effects of development proposals;
— To protect the productivity and capacity of natural systems and the ecological processes which maintain their functions;
— To promote development that is sustainable and optimizes resource use and management opportunities.

A comprehensive evaluation of all data associated with a site is performed. The data set includes available information on the reactor technology, planned and unplanned releases to the environment, the surrounding physical environment, the socioeconomic issues and data, etc. The resulting report may have different names in various countries (e.g. EIA report, environmental impact report, environmental impact statement or environmental statement). However, all are similar and comprehensively describe:

— The basis for the study;
— The data set (including the baseline environmental conditions at the selected site and a description of the selected technology or the PPE used if the technology is not selected yet);
— The analysis of the data;
— The description of the alternatives;
— An evaluation of the impacts of the alternatives during the construction, operation and decommissioning of the proposed NPP;
— Cumulative impacts, i.e. impacts that result when the environmental impacts associated with the proposed NPP are overlain on or added to impacts associated with other past, present and reasonably foreseeable future projects;
— A discussion of whether the impacts can be controlled, reversed or managed acceptably during the construction and operation of the proposed NPP.

The EIA identifies and analyses all environmental and socioeconomic impacts, including their nature, probability, duration (including permanence), magnitude and significance (whether positive or negative). It considers the entire project development programme, from construction through to decommissioning. Although the final EIA report is used for several purposes, its primary use is for decision makers to assess whether the proposed NPP project is environmentally acceptable at the selected site. If a fatal flaw, which cannot be corrected in practical terms by engineering design or by the application of mitigation measures, is discovered at this point in the process, it could lead to the site being disqualified. The EIA report presents measures to mitigate any identified significant impacts and provide a description and analysis of the steps or measures taken to be taken to reduce negative impacts as much as reasonably achievable or to compensate for the impacts to a level acceptable to the regulatory authorities who ultimately represent the interests of the stakeholders.

GSG-10 [15] provides recommendations and guidance on some of the characteristics of the radiological protection part of an EIA to be conducted at the early stage in the development of an NPP and the relation with the subsequent radiological environmental impact assessment to be conducted during the later authorization process by the nuclear regulatory body.

Once the EIA report has been approved by the competent authority, additional consideration of the possibility for adverse impacts is provided in the EMP for the construction and operation of the NPP (see Section 5.3.1).

5.2.2.1.   Preparation of the environmental impact assessment report

The preparation of the EIA report consists of several steps. Using the ESR as the basis, environmental monitoring and baseline information collection are fully implemented according to the approved protocols. The results form the basis for the assessment of impacts, which is also conducted according to the methodologies identified in the ESR. A complete EIA report includes consideration of both nuclear and non-nuclear concerns and impacts. In some countries, a non-nuclear assessment report may be developed to enable site preparation and other non-nuclear construction activities to commence. This would be followed by the full EIA. In such cases the authorization to commence site preparation and non-nuclear construction would contain provisions on the restoration of the site in the event the NPP project does not materialize.

The draft EIA report is submitted to the competent authority, e.g. the environmental regulatory body, to verify that the collected baseline data and the assessment methodologies are in line with the ESR and that, consequently, the EIA report adequately addresses the project issues. As mentioned in Section 4.3, the environmental and nuclear regulatory functions are often the responsibility of separate organizations. In such cases, the nuclear regulatory body will have an active role in the evaluation of the nuclear part of the EIA in coordination with the environmental regulatory body, and other relevant governmental agencies. The role of the nuclear regulatory body in an EIA is described in GSG-10 [15]. It may be that environmental monitoring highlights some issues that were previously unknown, and these will have to be addressed as well. Before the draft EIA report is ready for public comment, all of the issues raised by the regulatory body or discovered by the project proponent are expected to be addressed.

Good practice in stakeholder engagement suggests that the draft EIA report is also submitted to other key stakeholders for comment. Depending on the country, these stakeholders may include other organizations and agencies that support the competent authority, or public associations interested in the project. It is essential that the roles, responsibilities and authorities of the stakeholders are clearly defined so that their comments can be appropriately considered and addressed.

After all comments from the competent authority and other key stakeholders have been received and addressed, the final draft EIA report is submitted for comment to both the competent authority and the public, if such is stipulated in the country. The report could generate a significant public debate. Adjustments and amendments to the EIA report may be requested by stakeholders and the competent authority. Therefore, sufficient time for public consultation and potential amendments should be planned.

After all the comments on the draft EIA report have been received and considered by the preparers of the EIA report, the report is revised and reissued as a final draft EIA report to the competent environmental authority. The final approval by the competent authority represents the acceptance of the analyses and conclusions in the EIA report regarding the environmental impacts and environmental feasibility of the NPP project. This will need to be revised if conditions change, including design or operating parameters of the plant, change in environmental conditions, change in the assumptions used to prepare the EIA report, or other issues that may be identified as potentially impacting the conclusion of the EIA report.

The report will also identify the environmental sensitivities that would be addressed in the bid invitation specification, where site specific plant design provisions or construction techniques may be necessary to focus on these sensitivities. The report will further be used in the authorization process as well as detailed planning for the required environmental monitoring during construction and operation.

Depending on the country's legal and regulatory framework, the final approval may be appealed by any of the parties involved in the process, including stakeholders. The appeal process can result in significant time delays and extra costs if it is necessary to provide additional studies or information.

5.2.2.2.   Typical contents of an environmental impact assessment report for a nuclear power plant

The content of an EIA report for an NPP can to a large extent be considered to be similar to that of any other large industrial project's EIA, although there is variation depending on the project and the country. There are additional requirements for site studies and construction practice for an NPP as well as wider public interest which necessitates that the EIA report be very comprehensive. The report will need to consider radiological impacts, including transboundary impacts.

The EIA begins with a non-technical executive summary which includes a general description of the project and its justification, EIA procedure, environmental issues considered, magnitude and probability of significant impacts during construction and operation (normal and irregular), principles for decommissioning and mitigation measures including reference to the EMP. The summary also provides conclusions which serve as inputs to decision makers and planners, as well as for the broader public. The non-technical executive summary is likely to be widely reviewed by stakeholders, and therefore it is expected that it be comprehensive and written in a language understandable by the public. The non-technical executive summary is frequently translated into local languages.

To provide a context for the EIA, an introductory section describes the background of the proposed project, objective of the EIA, scope, national and international legislative framework, and the relationship of the document with the nuclear power authorization and decision making processes.

The procedures used for conducting the EIA as well as the stakeholder engagement plan, including the stages at which stakeholders are involved, are described.

The report further contains a detailed description of the project, the responsible party, project alternatives considered (including the no go option) and the project timetable from pre-construction through to the construction, commissioning, operation and decommissioning stages, recognizing that a new, more detailed EIA will be required for decommissioning.

As part of the description of the plant, a list of issues which have a bearing on environmental impact concerns are prepared, such as radiological and non-radiological emissions (both atmospheric and liquid); water and waste issues; chemicals potentially to be used on-site; transportation and traffic connections; and, if appropriate, a description of the nuclear fuel supply and management of spent nuclear fuel. The level of detail required varies depending on the national legislation and other requirements of the country. For example, natural hazards that are not known within the country, such as volcanism, may be addressed only briefly to assure the reader that the hazard was considered and dismissed as not relevant to the project.

The description of the environment would include details of the present condition of key affected environmental components as baseline information for future analysis, including primarily such topics as:

— Meteorology and air quality;
— Land use, buildings and land use plans;
— Soils, geology and hydrogeology;
— Water resources and quality;
— Terrestrial and wetland ecology;
— Aquatic ecology;
— Impact of climate change;
— Existing radiological and non-radiological contamination;
— Socioeconomic characteristics of the areas that may be affected by the NPP project;
— Historic and cultural resources.

A major component of the report is the description of the specialist assessment of the impact on different aspects of the environment. A description of the analysis used to estimate the magnitude and important characteristics of the impact during construction and operation, as well as in the scope of decommissioning of the proposed NPP, would also be included.

(a) Impacts during construction: A description of direct significant impacts of construction work on soil, bedrock, groundwater, archaeological finds, flora and fauna, land use and landscape, noise, air quality, people and society. The activities that might cause potential impacts include, but are not limited to:
    — Mobilization of material;
    — Human resources (direct and indirect employment);

— Expenditures by the NPP owners and the employees;
— Economic activity in the region due to increased employment and expenditures;
— Construction of the base camp;
— Construction of access roads;
— Provision of water for construction;
— Management of waste;
— Workshop building;
— Land preparation, land cut and fill;
— Transportation of project material;
— Construction of NPP buildings;
— Access to the site and transport of people.

(b) Impacts during normal operation of the NPP: A description of direct and indirect impacts on people and the environment due to plant operation. The activities that might create a potential impact include the additional workforce at the nuclear reactor and the operational activities themselves, which could generate heat release, radioactive releases to air and water, radioactive solid waste generation, generation of spent nuclear fuel, chemical and biological material release from cooling towers, chemical material release from laboratories, sanitation waste, activities related to operation and maintenance of switchyard and transmission lines leaving the site, and general maintenance activities.

(c) Impacts during accidents: A description of direct and indirect impacts on people and the environment due to unplanned releases. The factors that influence the magnitude of these impacts would include the following: the type of accident (design basis accident or design extension condition); source terms; the probability of occurrence of the accident; site specific meteorology, surface water and groundwater hydrology and hydrogeology; site specific demography; land and water use; and receptor habit data. If needed, the availability and implementation of EPR activities would also be a factor in the final impacts encountered by the people.

(d) Impacts during decommissioning: A description of the impact of the project on the environment during the decommissioning period. Some of the activities that might create a potential impact include the removal of spent fuel from the reactor core, dismantling of reactor components and the decontamination process. Although decommissioning will take place a substantial amount of time in the future and there is a general expectation that improvements in decommissioning techniques will be made over the intervening years, an analysis should be made in enough detail to show that decommissioning has been considered and can be performed using technologies available at the time of the analysis, resulting in an acceptable environmental impact so that the final site conditions will be consistent with the strategic plan (e.g. prepared for future use, returned to its existing state). Decommissioning requires a separate detailed EIA for activities after the end of the operational life of the plant. Since decommissioning is likely to commence many decades into the future, after 60–80 years of operation, the consideration of decommissioning conducted as part of the project EIA discussed here can only be a macro level review. Considerations at the design and construction phase that later influence decommissioning and the various approaches that are hence stipulated may be found in IAEA-TECDOC-1657, Design Lessons Drawn from the Decommissioning of Nuclear Facilities [45].

Specific impacts on aspects of the environment should be addressed in each section, including, but not limited to, impacts on:

— Air, soil and water quality due to nuclear and non-nuclear releases to the environment: This section provides a description of radioactive and non-radioactive emissions (during normal and abnormal operation) as well as other emissions (emergency power, heat generation and transportation). Guidance and information for the modelling of normal and abnormal operation radioactive release and its impact on people and the environment is provided in GSG-10 [15], NS-G-3.2 [16], Safety

Reports Series No. 19, Generic Models for Use in Assessing the Impact of Discharges of Radioactive Substances to the Environment [18], IAEA-TECDOC-1678, Environmental Modelling for Radiation Safety (EMRAS) – A Summary Report of the Results of the EMRAS Programme (2003–2007) [46], and Technical Reports Series No. 472, Handbook of Parameter Values for the Prediction of Radionuclide Transfer in Terrestrial and Freshwater Environments [47].

— Aquatic flora, fauna and ecological values of the area, including:
  • Conventional contaminant levels;
  • Aquatic biota populations and structure;
  • Ecological state;
  • Impact of discharge channels on habitats;
  • Impact of cooling water on water temperatures (cooling water modelling) and local ecology.
— Terrestrial flora, fauna and ecological values of the area, including:
  • Conventional contaminant levels;
  • Terrestrial biota populations and structure;
  • Ecological state.
— Wetlands flora, fauna and ecological values of the area, including:
  • Conventional contaminant levels;
  • Wetlands biota populations and structure;
  • Ecological state;
  • Impacts due to construction and operation of the NPP.
— Visual impact on the landscape and possible cultural environment values.
— Traffic amount, type and safety, including construction traffic and workforce traffic during operation and maintenance events.
— Noise level around the site area of the NPP operation and maintenance activities.
— People and socioeconomic factors, including:
  • Agriculture;
  • Fishing and hunting activities, both private and industrial;
  • Tourism;
  • Regional structure;
  • Economy and employment;
  • Changes in living conditions due to influx of temporary or permanent workers;
  • Requirement for additional infrastructure such as housing, police protection, education and health care;
  • Additional taxes to be paid to the area;
  • Potential health impacts (ionizing radiation and conventional contaminant related).
— Waste management: This section describes the amount of radiological and non-radiological waste and impacts of the waste management programmes. For the radiological waste management programme, actions should be presented in accordance with the radioactive waste classification and waste minimization principles [48–51].
— Spent fuel management: Although this is an issue that only needs to be fully addressed much later, after the start of the NPP operation, the description of the general concept considered for the spent fuel management would be the minimum to be included in the EIA, together with a timeline to be followed in fully addressing the issue.

Cumulative impacts are described, including those resulting from the combined impacts of the NPP project in addition to impacts from existing infrastructure and other past, present and reasonably foreseeable future projects. In addition, the cumulative impact over time on environmental resources that continue to be affected is described (e.g. water and air).

The impact of anticipated operational occurrences and accidents at the NPP on people and the environment due to design base accidents and design extension conditions at the NPP would be described. The area of impact and measures to address these impacts in case of accidents are also to be included.

Transboundary impacts will also need to be considered (e.g. impact from normal operation, impacts of accident conditions, socioeconomic impacts such as employment, and impacts on a shared watercourse).

As mentioned in Section 2.3.12, the national environmental legal and regulatory framework would provide guidance on the level of detail required to address nuclear fuel production activities that take place in another country. Typically, such activities are only described in the EIA report.

A detailed discussion would be provided regarding the prevention and mitigation of adverse impacts, including the measures to prevent and to diminish significant adverse impacts of the project. The selection criteria for the proposed mitigation measures, for example, cost, technical feasibility, legal possibility or social acceptability, is clarified in this section. The hierarchy of possible approaches for the mitigation of environmental impacts is presented. Examples typically include:

— Engineering and planning alterations;
— Practice alterations for construction and operation;
— Habitat restoration;
— Financial compensation;
— Communication of information or other measures found acceptable by the country.

A description of the EMP for the construction and operation periods would be provided. Necessary monitoring within this framework will be based on the results of the EIA and is generally established for the environment that may be affected, such as groundwater, surface water, soil or biota.

5.2.2.3. Impact evaluation methodology

A clear description of the scope of the EIA is necessary to provide a background for the selection of the assessment methodology, its application and the identification of uncertainties. For the spatial boundaries, the customary approach is to assess at a local level (on-site and immediately off the site) and at a regional level (as far as impacts are likely to be observed, which varies depending on different environmental aspects). The temporal boundaries take into account the long operational life and possible life extension of the NPP, as well as the presence of radioactivity beyond plant operation. Details of this scope and methodological approach would be identified in the ESR and agreed with the competent authority prior to proceeding further with the development of the EIA report.

The definition of what constitutes a significant impact to be avoided or minimized would be identified in the ESR and is an important early step for the competent authority. The methodology to evaluate the significance of an environmental impact could be based on qualitative, quantitative and perhaps intangible parameters of the impact. Some possible examples include:

— Comparison of the impact to applicable regulatory limits and international standards;
— Expert opinion;
— Probability;
— Reversibility;
— Extent: geographical and in population;
— Intensity;
— Duration;
— Uncertainty;
— Cumulative impact with time, together with the impact from sources other than the project in question.

Impacts that are found to be significant and adverse would undergo planning for avoidance or mitigation, and would be addressed in the monitoring plan, although some impacts need not be actively monitored. Beneficial impacts are expected to be identified as well thus providing a holistic evaluation of the project. An overall cost–benefit analysis may be performed to evaluate the economic value of the entire project.

5.2.2.4. Use of models in environmental impact assessments

In some situations, expert judgement is the only available means of assessment of the impact of an environmental aspect. However, for complex projects such as NPP projects, expert judgement is used as a sufficient argument only in specific topic areas where modelling is not possible or not yet developed, e.g. for ecological impacts from construction related activities on wetlands and threatened and endangered species. There are various calculation methods and modelling approaches that have shown good results in impact prediction and can be used to identify the effect of the impacts on the environment. Table 1 provides some examples of impact model approaches used in nuclear EIA reports.

In some countries, the competent authority may recommend a particular model for use in the EIA report. However, it is the responsibility of the EIA team to choose the model and calculation approach, provided that the team has satisfied the competent authority with regard to the model's suitability and accuracy.

Consideration of radiological impacts requires specific modelling approaches. A common model approach to estimate the release of radionuclides to the environment is using atmospheric dispersion and dose modelling (see Safety Reports Series No. 19, Generic Models for Use in Assessing the Impact of Discharges of Radioactive Substances to the Environment [18] and Technical Reports Series No. 472, Handbook of Parameter Values for the Prediction of Radionuclide Transfer in Terrestrial and Freshwater Environments [47]). The approach requires data on source terms received at an early stage from the technology vendor, as well as knowledge of the pathways and receptors of the radionuclides transmitted via aerodynamic transport. The output of this approach is calculation of the transport of various radionuclides released in the atmosphere and the estimation of doses received by people and the environment.

Planned discharges to the atmosphere of airborne effluents and to surface aquatic media of liquid effluents do occur during normal operation of the NPP. Although plants are designed to prevent accidental releases, the EIA should consider that such releases could occur and impact humans and the environment. This is primarily handled in the development of the EMP, since, by definition, an accident that can be reasonably foreseen has to be addressed in the NPP design in order to prevent its occurrence [52]. Figure 4 shows the potential exposure pathways of radionuclide release to the public and the environment [42].

5.2.3. Addressing uncertainties in the environmental impact assessment process

Uncertainties in the information available to assess and analyse the potential of nuclear power in a country will lead to uncertainties in the calculation of its environmental impact, necessitating additional evaluation as new information is collected in the process. The level of acceptable uncertainty on the impact will depend on the irreversibility of the decision being made at that time, and the level of risk of the project ultimately being deemed unacceptable or prohibitively expensive (i.e. not cost effective as a source of energy). For example, proceeding to the next stage of a desktop study could accommodate a high level of uncertainty as only the project proponent's time and resources are at risk, whereas proceeding to site construction activities will impact the environment and necessitate the expenditure of significant resources. Thus, the risk of requiring significant changes to a project is minimized early in the process through the collection of sufficient data and thorough discussions with all stakeholders.

The initial studies in Phase 1 also seek to identify the environmental factors that would immediately exclude a region or potential sites from further consideration. These could include established protected areas, sites with confirmed presence of rare, threatened or endangered species, existing archaeological or historical protected areas, and recently obtained demographic characteristics. In cases where there is uncertainty regarding the environment's status, potential areas or sites could be included in a list pending further investigation.

A full scoping and EIA process are undertaken on the selected site(s) to obtain more detailed information and analyse site specific data to resolve uncertainties about the site. A more detailed discussion of addressing uncertainties during the EIA process is included in Annex II.

TABLE 1. SOME IMPACT MODEL APPROACHES USED IN NUCLEAR ENVIRONMENTAL IMPACT ASSESSMENT REPORTS

Impact	Model approach	Disciplines involved	Input	Output	Model specifics
Radiological	Transport and dose modelling[a]	Defining source terms Dispersion modelling Radioactive decay products Radioecology Dose calculations	Source term Pathways[b] Impact receivers Habit data for receivers	Dispersion Estimated concentrations Estimated doses	Due to its importance, approval of the model should be granted by the competent authority and may require more time.
Thermal and water use	Mathematical dispersion modelling	Dispersion modelling Hydrological modelling depending on water body used for cooling	Discharge temperature and flow rates, location and technology, details of water body receiving the heat discharge	Assessment of impacts on water temperatures and thermal stratification	Model calibration may require time consuming measurements.
Aquatic biota and their life cycles, thermal impact, impingement, entrainment, changes in the ecosystem	Demographic approaches	Hydrobiology Oceanography Fisheries, hydrodynamic and water quality modelling	Aquatic field studies Species population data Ecosystem structure Results from thermal modelling	Impacts on aquatic species Biodiversity changes	Aquatic field studies through the seasonal variations may be very laborious.
Terrestrial, wetlands, marine and freshwater biota (radiological risk)	Integrated exposure/dose/ effect assessment with risk characterization	Radioecology	Environmental concentrations Dose conversion coefficients Concentration ratios Distribution coefficients	Dose rates Risk quotients	Tiered approach Selected animals and plants
Regional economy (employment, revenues, economy of the region)	Economic modelling	Economic modelling, analysis and projections	Regional economy data Development plans Project financial parameters	Prediction of the economic impacts on the region	Financial risks may influence the model output.

TABLE 1. SOME IMPACT MODEL APPROACHES USED IN NUCLEAR ENVIRONMENTAL IMPACT ASSESSMENT REPORTS (cont.)

Impact	Model approach	Disciplines involved	Input	Output	Model specifics
People and society (migration, quality of life, culture, environmental justice)	Expert opinion Social trend projections Computer modelling	Socioeconomic impact assessment Communications Resident surveys Media analysis	Socioeconomic characteristics of the population Personal perceptions, opinions and fears	Social impacts from migration and revenue change Environmental justice and cultural heritage considerations Increased communication	Needs to be carefully coordinated with stakeholder engagement processes.

a   For more information on the types of model, see Refs [18, 46].
b   Data from baseline meteorology, oceanography, hydrology and groundwater surveys in the investigated area.

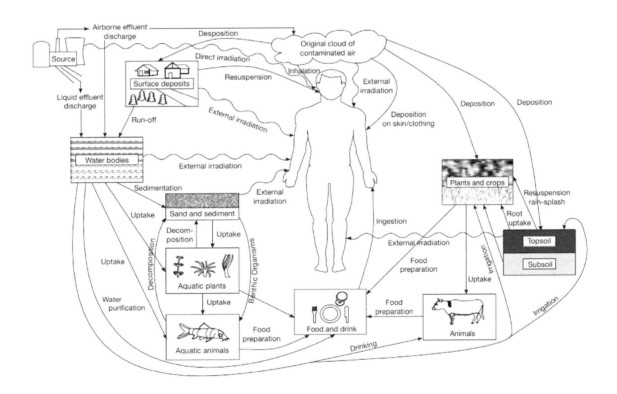

FIG. 4. The possible pathways of exposure for members of the public as a result of releases of radioactive material to the environment [42].

During construction of the NPP, unexpected environmental conditions may be identified that need to be resolved either by engineering or administrative controls, or, rarely, by changing the location or certain key features of the NPP itself. For example, the stack height may need to be changed, or a planned effluent discharge location may need to be moved. The earlier these decisions can be made in the programme, the smaller their expected impact on budget and schedule.

Conclusions about the impact an NPP project will have on the environment involve the collection and analysis of environmental data as well as a thorough understanding of the reactor technology, particularly its potential releases to the environment. Additionally, the interactions between the technology and the environment are expected to be known and fully understood. Since the plant life cycle of the plant comprises decades, estimates are made of the future environmental conditions, technology and interactions. All of this contains uncertainties, and the EIA process seeks to understand where uncertainties exist and to reduce the uncertainties as much as possible.

During the development of the EIA, the technology may not yet be selected and the initial assessments may require some bounding assumptions or use of generic data for both the environment and the technology in order to allow an initial assessment to be made according to the required overall schedule of the development programme.

5.2.4. Bid invitation specification

Apart from its primary purpose of supporting the application for an environmental authorization for the proposed project, the EIA report (including specification of the site environmental conditions, characteristics and data) and specifically the EMP provide key information to include in a bid invitation or to guide negotiations with a preferred vendor.

Site conditions have a great influence on the layout, design, construction and costs of the NPP. Comprehensive specification of environmental site conditions, factors, characteristics and data, including those that may seem not to be directly related to the project, are provided in the bid invitation specification or included in negotiations with a preferred vendor in as much detail as possible.

The owner typically offers bidders or the preferred vendor open access to all detailed site studies, including EIA documents, the EMP and collected site data. The owner also establishes a procedure for the resolution of vendors' questions regarding the interpretation of the site data and the matters mentioned above. In the event that the EIA report is completed before a bid invitation specification or guideline for negotiation with a preferred vendor is issued, the owner would include in the bid or guideline all commitments, limitations and conditions resulting from the EIA report approval by the competent authority. These commitments, limitations and conditions would be described in detail in the EMP. However, if the report is not completed and approved by the competent authority, the owner is expected to ensure that these commitments, limitations and conditions are communicated to a vendor at least at the contract negotiation stage and incorporated in a contract. Otherwise, project risk is increased, and the owner can face unforeseen expenses during the course of project implementation. Furthermore, the contract typically has mechanisms in place to resolve future possible authorization issues and conditions to enable continuous project implementation.

If a competitive bidding process is being followed, or negotiations are being conducted with a preferred vendor, the competing bids (or the offer from the preferred vendor) are judged (evaluated), among other factors, on the basis of environmental impacts from the proposed NPP. This means that bids are evaluated, at a minimum, as to whether they comply with the results of the EIA and any conditions specified in the environmental authorization.

Once the reactor technology has been selected, there may still be decisions to be made on aspects such as the reactor chemistry regimes and waste management that may affect the environmental impacts and discharge authorizations. These will be part of the activities performed in Phase 3 and addressed through the EMP.

5.3. ENVIRONMENTAL PROTECTION ACTIVITIES IN PHASE 3 AND BEYOND

5.3.1. Environmental management plan

The EIA process identifies environmental impacts resulting from the construction, operation and decommissioning of an NPP. The EMP is a tool to ensure that mitigation measures identified during the EIA are documented and executed throughout the project life cycle. The EMP is developed during the EIA process and generally submitted with the EIA report. In some instances, the EMP (or a draft EMP) is approved as part of the environmental authorizations or record of decision. However, if there are uncertainties around aspects such as technology selection, the draft EMP will need to be revised and approved by the competent authority prior to construction, allowing for the integration of the detailed design into planning for construction.

5.3.1.1. *Objectives of the environmental management plan*

The objective of the EMP is to ensure that management and mitigation actions arising from the EIA are identified, described and executed during the life cycle of the project. The process for the EMP provides an opportunity for stakeholders to contribute to, and where appropriate, to continually be involved with the environmental management of the project. In this context, stakeholders include the competent authority, an environmental monitoring committee (if formally established), local communities, non-governmental organizations and funders. The EMP serves as a reference document which can be used for all monitoring including that of on-site, daily activities. The EMP would link to and be enhanced with an environmental management system, generally during operation, but can also be introduced during construction and assist with transitional phases from construction to operation, and from operation to the EIA for decommissioning.

5.3.1.2. *Contents of the environmental management plan*

During the EIA process some information may be less detailed at the time the EIA report is submitted to the authorities for approval. A revised EIA is prepared when more detail becomes available. This will also impact on the level of detail that can be included in the EMP. The level of uncertainty and information available during the EIA process such as the selection of technology and detailed design can be addressed through the development of a framework EMP which is further developed and detailed as the project progresses through the bidding process and into construction. This approach provides sufficient information to demonstrate commitment to the authorities, informs the bidding process and satisfies other interested and affected parties, including lenders.

The EMP will be based on information available during the execution of the EIA, including the ESR, the EIA report and specialist studies conducted as part of scoping and EIA. All phases of the project from pre-construction to decommissioning require an EMP but it is advised to separate the EMPs for construction, operation and decommissioning. The construction EMP typically includes the following elements:

— Background information on the project, including a description of the activities to be conducted during different phases of the project;
— Information on the proponent and their commitment to management actions to mitigate negative environmental impacts and to enhance positive impacts;
— A description of the receiving environment with respect to biophysical, economic and social aspects;
— How the EMP will integrate into the project and link with other matters such as risk, health, safety and quality management;
— Applicable legal requirements including acts, regulations, guidelines and standards;
— Applicable authorizations and at what point these will be required;
— The proponent's current policies and procedures and their linkage to the construction EMP;
— Predicted positive and negative impacts and recommendations/management actions stipulated in the EIA report;
— Roles and responsibilities of key role players such as the proponent, contractors, subcontractors, local and national authorities and monitoring committees;

— The implementation programme, which links directly to the management actions identified during the preceding activities. The implementation programme provides the detailed actions and is generally broken down into the following categories:
  • Environmental impacts identified during the EIA requiring management;
  • Ongoing environmental monitoring;
  • Detailed management action;
  • Party responsible for the action;
  • Criteria and targets;
  • Dates and times for completion of actions;
  • Reporting requirements;
  • Auditing procedures and activities that will provide an independent statement of compliance.

To improve the effectiveness of the EMP, the process to communicate the requirements of the EMP and provide the necessary awareness and training should be described. This training and awareness typically includes the content and purpose of the EMP, content of method statements, monitoring and performance reports, incident reporting and complaints.

The EMP is further enhanced by the use of a method statement which is developed by the contractor or subcontractor to clearly specify how the identified environmental impact will be managed. The statement includes criteria against which success can be measured.

The EMP is a living document that can be reviewed and adapted when required. Periodic safety reviews performed for the NPP may identify changes in the state of knowledge of impacts of hazards, regulations and the local environment and may result in an updating of the EMP. However, a governance process is necessary to effect these changes. If the proposed changes are significant, it may be necessary for the environmental monitoring committee (if such a committee is formally established) to support the change and for the competent authority to approve it, depending on the governance model of the NPP project and applicable legislation of the country.

5.3.2. Environmental monitoring programme

A component of the EMP is the environmental monitoring programme, a tool which checks that the environmental impact during construction and operation stays within assessed and accepted limits and, in case it does not, specifies a process to address the activity causing the observed exceedance values.

Typically, the environmental monitoring programme is based on:

— Baseline information collected from Phase 1 and throughout the EIA report preparation;
— Considerations related to the selected technology (as sources of impacts);
— Specialist studies and the results of the EIA report;
— Application for and conditions in all authorizations.

The environmental monitoring programme checks that that significant environmental impacts are monitored to provide timely assurance that they are within the authorization limits. The environmental monitoring programme scope would be aligned with the level of risk and/or uncertainty identified during the EIA process and as construction is initiated. Some risks, such as the extent of archaeological finds, may not be known prior to construction. In terms of releases, a methodology is expected to be in place to identify potential release points early enough to activate a more detailed assessment of a release. Typical examples of monitoring activities include:

— Measurements of concentration in air (emissions);
— Groundwater (wells, etc.) collection;
— Surface water sampling;
— Soil erosion;

— Temperature measurements for the affected water bodies;
— Bioindicator sampling;
— Socioeconomic issues;
— Archaeological finds and preservation.

The environmental monitoring programme includes a definition of required data reliability, the frequency of data collection, monitoring and sampling locations (along the exposure pathways) and the density of these locations. Monitoring points will be on the site as well as off the site. Additionally, the monitoring programme may expand and include different activities during the life cycle of the NPP.

Roles and responsibilities are detailed in the EMP. The plant owner or operator is responsible for ensuring that all legal environmental protection legislative framework requirements, including implementation of the environmental monitoring programme, are fulfilled. The environmental management team on-site would have the appropriate level of authority to ensure compliance with the EMP, which will also ensure no unnecessary delays due to environmental issues.

Environmental agencies and/or the competent authority perform surveillance, auditing, independent monitoring and other defined activities according to an environmental protection legislative framework to ensure that the owner follows all prescribed duties. In some countries, the environmental authority may require the owner to appoint an independent environmental control officer whose responsibility it is to monitor compliance and report on a regular, stipulated timeline. As indicated above, it may also be a requirement to appoint an environmental compliance committee consisting of various stakeholders, including the authorities and community representatives. Such a committee would also have an independent chairperson. If the authorities do not have the resources to appoint an auditor, they may request that the owner appoints an independent auditor. The audit results would then be submitted to the authority.

An NPP project specifically requires a radiological environmental monitoring programme which starts prior to construction. The baseline environmental data collection programme to support the EIA should include the same elements and sampling points. Guidance on radiological monitoring programmes and activities, including techniques, procedures and data interpretation and evaluation, can be found in NS-G-3.2 [16], RS-G-1.8 [42] and Safety Reports Series No. 67, Monitoring for Compliance with Exemption and Clearance Levels [53].

The environmental monitoring programme, especially its radiological component, is usually designed for normal operations. However, arrangements for emergency preparedness are expected to be considered carefully when implementing the monitoring programme during the pre-operational stage. The basic intervention levels are developed and understood by all responsible persons and organizations involved in the emergency response. The operational intervention levels refer to parameters that can be easily measured so that an interpretation can be made rapidly if intervention is required. During operation, if there is an abnormal occurrence, specific monitoring would be developed and modified to adapt to the status of the event (ongoing, ended, pre-remediation and post-remediation). The specific objectives of the ongoing event, or the emergency radiation monitoring in the environment, are:

— To provide baseline data and information prior to operations;
— To provide accurate and timely data on the level and degree of hazards resulting from a radiation emergency, in particular, on the levels of radiation and environmental contamination with radionuclides;
— To assist decision makers on the need to make interventions and take protective actions;
— To provide information for the protection of emergency workers;
— To provide information to the public on the degree of hazard;
— To provide information required to identify any people for whom long term medical screening is warranted.

5.3.3. Environmental management activities after operations cease

When an NPP reaches the end of its operating life, a number of complex activities commence to stabilize the plant, remove the radioactive components, and prepare the plant for decommissioning to return the site to as close to its original condition as possible or to the condition specified in the strategic plan, such as for a future use. When the decommissioning plan is established, a new EIA is developed based on the environmental conditions currently experienced, the current regulatory requirements and new stakeholder input.

In general, environmental impacts will be of two different types: (a) radiological releases and (b) environmental impacts due to the major construction project involved. Long before the operating life is completed, discussions would be held with regulators and stakeholders as to what is expected, and an environmental decommissioning plan developed. Knowing the expectations of the decommissioning plan early in the nuclear power programme will enable treatment of the plant and its effluents to be handled so as to minimize impacts after operations cease [54].

5.3.4. Environmental performance review

During construction and operation, it is necessary to review the overall environmental performance. This is in addition to the continuous oversight carried out by the owner/operator, and the environmental monitoring committee (if such a committee is formally established). This review may occur at intervals from every several years to more frequently as required. Various stakeholders, including the authorities, interested and affected communities, and lenders, could also influence the frequency and scope of such environmental performance reviews.

6. CONCLUSION

Although environmental protection is identified as a separate infrastructure issue in the Milestones approach [2], it is related to the majority of the other issues. As such, it is necessary that its development and implementation be coordinated with other related infrastructure issues.

Environmental protection for nuclear power programmes has both radiological and non-radiological aspects. Radiological aspects are primarily related to exposure to direct radiation and exposure to radionuclides released to the environment from the NPP during its operation (whether routine or under accident conditions) and decommissioning, and from the facilities used to manage the radioactive waste and spent nuclear fuel generated by the NPP. Non-radiological impacts are similar to those that occur for other, non-nuclear power production facilities.

Activities related to environmental protection for nuclear power programmes span across all three phases of the Milestones approach [2] and continue during the operational life of the NPP through to its eventual decommissioning.

Stakeholder involvement and public communication play a crucial role in environmental protection for nuclear power programmes. Therefore, this component of the nuclear power programme is expected to be carefully planned and integrated into the environmental protection programme.

There are many environmental authorizations needed during the development and implementation of a nuclear power programme. They are issued by different national, regional or local authorities; thus, planning and coordination among all authorities involved is an important component of the environmental protection programme. Appropriate agreements would typically be implemented between the national regulatory bodies to ensure the coordination and consistency of their respective regulatory functions and activities.

It is important to have clear and enforceable laws and regulations so that all relevant authorities, the NEPIO, the regulatory body or bodies, and the owner/operator, understand their roles and responsibilities and the interface with each another. A well-defined mechanism for resolution of any conflicts that arise is needed.

The skill set needed to successfully implement the environmental protection programmes for nuclear power programmes include both the traditional, non-radiological disciplines, and the disciplines related to radiological impact and its assessment. Careful planning at the beginning of the nuclear power programme is needed to ensure the availability of staff with the required skill level in the different functional areas.

If the proposed NPP is close to an international border or if the region of influence of the NPP (in particular with regard to emergency planning) is likely to extend into the area of another country, arrangements would be made among the countries involved to share information and to accurately assess the potential radiological (and, if appropriate, non-radiological) environmental impact of the proposed NPP.

The EIA process identifies environmental impacts resulting from construction and operation.[4] An EMP would accompany the EIA report and guide the environmental protection activities before and during construction and operation. An environmental monitoring programme is expected to be put in place to ensure that the conditions set out in the EMP are implemented as required, and to determine whether environmental quality at the NPP and in its vicinity of conforms to defined criteria. Both the EMP and the monitoring programme would be updated if there are major changes in the design of the NPP or in environmental conditions or if the monitoring programme results indicate that a different strategy or emphasis is required.

REFERENCES

[1]   INTERNATIONAL ATOMIC ENERGY AGENCY, Nuclear Energy Basic Principles, IAEA Nuclear Energy Series No. NE-BP, IAEA, Vienna (2008).

[2]   INTERNATIONAL ATOMIC ENERGY AGENCY, Milestones in the Development of a National Infrastructure for Nuclear Power, IAEA Nuclear Energy Series No. NG-G-3.1 (Rev. 2) (in press).

[3]   INTERNATIONAL ATOMIC ENERGY AGENCY, Evaluation of the Status of National Nuclear Infrastructure Development, IAEA Nuclear Energy Series No. NG-T-3.2 (Rev. 2), IAEA, Vienna (2022).

[4]   INTERNATIONAL ATOMIC ENERGY AGENCY, Responsibilities and Functions of a Nuclear Energy Programme Implementing Organization, IAEA Nuclear Energy Series No. NG-T-3.6 (Rev. 1), IAEA, Vienna (2019).

[5]   INTERNATIONAL ATOMIC ENERGY AGENCY, Initiating Nuclear Power Programmes: Responsibilities and Capabilities of Owners and Operators, IAEA Nuclear Energy Series No. NG-T-3.1 (Rev. 1), IAEA, Vienna (2020).

[6]   INTERNATIONAL ATOMIC ENERGY AGENCY, Nuclear Infrastructure Bibliography, https://www.iaea.org/topics/infrastructure-development/bibliography

[7]   INTERNATIONAL ATOMIC ENERGY AGENCY, Strategic Environmental Assessment for Nuclear Power Programmes: Guidelines, IAEA Nuclear Energy Series No. NG-T-3.17, IAEA, Vienna (2018).

[8]   INTERNATIONAL ATOMIC ENERGY AGENCY, Managing Siting Activities for Nuclear Power Plants, IAEA Nuclear Energy Series No. NG-T-3.7 (Rev. 1), IAEA, Vienna (2022).

[9]   INTERNATIONAL ATOMIC ENERGY AGENCY, Site Survey and Site Selection for Nuclear Installations, IAEA Safety Standards Series No. SSG-35, IAEA, Vienna (2015).

[10]   INTERNATIONAL ATOMIC ENERGY AGENCY, Building a National Position for a New Nuclear Power Programme, IAEA Nuclear Energy Series No. NG-T-3.14, IAEA, Vienna (2016).

[11]   INTERNATIONAL ORGANIZATION FOR STANDARDIZATION, Environmental Management Systems - Requirements with Guidance for Use, ISO 14001, ISO, Geneva (2015).

---

[4] The EIA process for a proposed NPP would include a qualitative discussion of the likely impacts during the future decommissioning of the NPP, but a separate EIA would be required for the actual decommissioning process prior to it being undertaken.

[12] INTERNATIONAL ATOMIC ENERGY AGENCY, Stakeholder Engagement in Nuclear Programmes, IAEA Nuclear Energy Series No. NG-G-5.1, IAEA, Vienna (2021).

[13] INTERNATIONAL ATOMIC ENERGY AGENCY, Communication and Consultation with Interested Parties by the Regulatory Body, IAEA Safety Standards Series No. GSG-6, IAEA, Vienna (2017).

[14] INTERNATIONAL ATOMIC ENERGY AGENCY, Site Evaluation for Nuclear Installations, IAEA Safety Standards Series No. SSR-1, IAEA, Vienna (2019).

[15] INTERNATIONAL ATOMIC ENERGY AGENCY, Prospective Radiological Environmental Impact Assessment for Facilities and Activities, IAEA Safety Standards Series No. GSG-10, IAEA, Vienna (2018).

[16] INTERNATIONAL ATOMIC ENERGY AGENCY, Dispersion of Radioactive Material in Air and Water and Consideration of Population Distribution in Site Evaluation for Nuclear Power Plants, IAEA Safety Standards Series No. NS-G-3.2, IAEA, Vienna (2002).

[17] INTERNATIONAL ATOMIC ENERGY AGENCY, Hazards Associated with Human Induced External Events in Site Evaluation for Nuclear Installations, IAEA Safety Standards Series No. SSG-79, IAEA, Vienna (2023).

[18] INTERNATIONAL ATOMIC ENERGY AGENCY, Generic Models for Use in Assessing the Impact of Discharges of Radioactive Substances to the Environment, Safety Reports Series No. 19, IAEA Vienna (2001).

[19] INTERNATIONAL ATOMIC ENERGY AGENCY, WORLD METEOROLOGICAL ORGANIZATION, Meteorological and Hydrological Hazards in Site Evaluation for Nuclear Installations, IAEA Safety Standards Series No. SSG-18, IAEA, Vienna (2011)

[20] FOOD AND AGRICULTURE ORGANIZATION OF THE UNITED NATIONS, INTERNATIONAL ATOMIC ENERGY AGENCY, INTERNATIONAL CIVIL AVIATION ORGANIZATION, INTERNATIONAL LABOUR ORGANIZATION, INTERNATIONAL MARITIME ORGANIZATION, INTERPOL, OECD NUCLEAR ENERGY AGENCY, PAN AMERICAN HEALTH ORGANIZATION, PREPARATORY COMMISSION FOR THE COMPREHENSIVE NUCLEAR-TEST-BAN TREATY ORGANIZATION, UNITED NATIONS ENVIRONMENT PROGRAMME, UNITED NATIONS OFFICE FOR THE COORDINATION OF HUMANITARIAN AFFAIRS, WORLD HEALTH ORGANIZATION, WORLD METEOROLOGICAL ORGANIZATION, Preparedness and Response for a Nuclear or Radiological Emergency, IAEA Safety Standards Series No. GSR Part 7, IAEA, Vienna (2015).

[21] INTERNATIONAL ATOMIC ENERGY AGENCY, Policies and Strategies for Radioactive Waste Management, IAEA Nuclear Energy Series No. NW-G-1.1, IAEA, Vienna (2009).

[22] INTERNATIONAL ATOMIC ENERGY AGENCY, Policies and Strategies for the Decommissioning of Nuclear and Radiological Facilities, IAEA Nuclear Energy Series No. NW-G-2.1, IAEA, Vienna (2011).

[23] Convention on Access to Information, Public Participation in Decision-Making and Access to Justice in Environmental Matters (the Aarhus Convention), United Nations Economic Commission for Europe, Aarhus (1998).

[24] Convention on Environmental Impact Assessment in a Transboundary Context (the Espoo Convention), United Nations Economic Commission for Europe, Espoo (1991).

[25] Convention on Wetlands of International Importance especially as Waterfowl Habitat (the Ramsar Convention), United Nations Educational, Scientific and Cultural Organization, Ramsar (1971).

[26] Convention on the Prevention of Marine Pollution by Dumping of Wastes and Other Matter (the London Convention), International Maritime Organization, London (1974).

[27] Montreal Protocol on Substances That Deplete the Ozone Layer (the Montreal Protocol), United Nations Environment Programme, Montreal (1987).

[28] Basel Convention on the Control of Transboundary Movements of Hazardous Wastes and their Disposal (the Basel Convention), United Nations Environment Programme, Basel (1989).

[29] Bamako Convention on the Ban of the Import into Africa and the Control of Transboundary Movement and Management of Hazardous Wastes within Africa (the Bamako Convention), Organization of African Unity, Bamako (1991).

[30] Convention for the Protection of the Marine Environment of the North-East Atlantic (the OSPAR Convention), OSPAR Commission, Oslo (1992).

[31] United Nations Framework Convention on Climate Change, United Nations, New York (1992)

[32] Convention for the Protection of the Marine Environment and the Coastal Region of the Mediterranean (the Barcelona Convention), United Nations Environment Programme, Barcelona (1995).

[33] Paris Agreement under the United Nations Framework Convention on Climate Change (the Paris Agreement), United Nations Framework Convention on Climate Change, Paris (2015).

[34] Convention on Early Notification of a Nuclear Accident, INFCIRC/335, IAEA, Vienna (1986).

[35] Convention on Assistance in the Case of a Nuclear Accident or Radiological Emergency, INFCIRC/336, IAEA, Vienna (1986).

[36] Convention on Nuclear Safety, INFCIRC/449, IAEA, Vienna (1994).

[37] Joint Convention on the Safety of Spent Fuel Management and on the Safety of Radioactive Waste Management, INFCIRC/546, IAEA, Vienna (1997).

[38] Convention on Third Party Liability in the Field of Nuclear Energy (the 1960 Paris Convention), OECD Nuclear Energy Agency, Paris (1960).

[39] Vienna Convention on Civil Liability for Nuclear Damage, INFCIRC/500, IAEA, Vienna (1996).

[40] Convention on Supplementary Compensation for Nuclear Damage, INFCIRC/567, IAEA, Vienna (1998).

[41] EQUATOR PRINCIPLES ASSOCIATION, The Equator Principles (2020), https://equator-principles.com/wp-content/uploads/2021/02/The-Equator-Principles-July-2020.pdf

[42] INTERNATIONAL ATOMIC ENERGY AGENCY, Environmental and Source Monitoring for Purposes of Radiation Protection, IAEA Safety Standards Series No. RS-G-1.8, IAEA, Vienna (2005).

[43] INTERNATIONAL ATOMIC ENERGY AGENCY, Programmes and Systems for Source and Environmental Radiation Monitoring, Safety Reports Series No. 64, IAEA, Vienna (2010).

[44] INTERNATIONAL ASSOCIATION FOR IMPACT ASSESSMENT, Principles of Environmental Impact Assessment Best Practice, IAIA, Fargo, ND (1999).

[45] INTERNATIONAL ATOMIC ENERGY AGENCY, Design Lessons Drawn from the Decommissioning of Nuclear Facilities, IAEA-TECDOC-1657, IAEA, Vienna (2011).

[46] INTERNATIONAL ATOMIC ENERGY AGENCY, Environmental Modelling for Radiation Safety (EMRAS) — A Summary Report of the Results of the EMRAS Programme (2003–2007), IAEA-TECDOC-1678, IAEA, Vienna (2012).

[47] INTERNATIONAL ATOMIC ENERGY AGENCY, Handbook of Parameter Values for the Prediction of Radionuclide Transfer in Terrestrial and Freshwater Environments, Technical Reports Series No. 472, IAEA, Vienna (2010).

[48] INTERNATIONAL ATOMIC ENERGY AGENCY, Classification of Radioactive Waste, IAEA Safety Standards Series No. GSG-1, IAEA, Vienna (2009).

[49] INTERNATIONAL ATOMIC ENERGY AGENCY, Disposal Approaches for Long Lived Low and Intermediate Level Radioactive Waste, IAEA Nuclear Energy Series No. NW-T-1.20, IAEA, Vienna (2009).

[50] INTERNATIONAL ATOMIC ENERGY AGENCY, Predisposal Management of Radioactive Waste, IAEA Safety Standards Series No. GSR Part 5, IAEA, Vienna (2009).

[51] INTERNATIONAL ATOMIC ENERGY AGENCY, Disposal of Radioactive Waste, IAEA Safety Standards Series No. SSR-5, IAEA, Vienna (2011).

[52] INTERNATIONAL ATOMIC ENERGY AGENCY, Safety of Nuclear Power Plants: Design, IAEA Safety Standards Series No. SSR-2/1 (Rev. 1), IAEA, Vienna (2016).

[53] INTERNATIONAL ATOMIC ENERGY AGENCY, Monitoring for Compliance with Exemption and Clearance Levels, Safety Reports Series No. 67, IAEA, Vienna (2012).

[54] INTERNATIONAL ATOMIC ENERGY AGENCY, Decommissioning of Facilities, IAEA Safety Standards Series No. GSR Part 6, IAEA, Vienna (2014).

Annex I

EXPERTISE FOR ENVIRONMENTAL PROTECTION PROCESSES FOR NUCLEAR POWER

I–1. INTRODUCTION

An embarking country will need to ensure that there are sufficient human resources (both in number and expertise to perform both the assessment and its independent review on behalf of regulatory bodies), material assets (e.g. instrumentation), and analysis techniques (e.g. software) available to the project at the time they are needed. EIAs require a wide range of specialists. The role of a specialist in an EIA is to support the decision making process through the assessment of benefits and negative impacts of the project and to provide practical recommendations to enhance the benefits and mitigate the negative impacts.

I–2. NECESSARY SPECIALISTS

Specialists will be required with skills in the following fields of expertise, including modelling and statistical skills:

— Physical environment;
— Meteorology;
— Air quality;
— Land use;
— Terrestrial ecology (flora, fauna, vertebrates and invertebrates);
— Fresh water ecology;
— Marine ecology;
— Wetlands ecology;
— Geology/geotechnology/hydrogeology (seismic, volcanic hazards);
— Archaeology, palaeontology;
— Surface water and groundwater hydrology;
— Social sciences;
— Social impact (including demographical) assessment;
— Cultural and heritage;
— Economy;
— Agricultural economy, land use;
— Engineering;
— Traffic engineering;
— Town and regional planning;
— Acoustical engineering;
— Landscape architecture (visual impact assessment);
— Electromagnetic interference;
— Radioecology;
— Radiological protection;
— Radioactive waste;
— Emergency planning;
— Legal and authorization;
— Grid engineering;

— Project management, management systems, quality systems;
— Public participation and facilitation;
— Environmental justice;
— Stakeholder engagement;
— Security.

Table I–1 summarizes the typical areas of expertise and the associated activities related to environmental protection in each phase of the Milestones approach [2].

TABLE I–1. TYPICAL AREAS OF EXPERTISE AND ACTIVITIES REQUIRED FOR ENVIRONMENTAL PROTECTION IN EACH PHASE

Area of expertise	Phase 1	Phase 2	Phase 3
Meteorology and air quality modelling	Incorporate available information into the baseline information.	Start collecting regional historic data for hazard and environmental analysis; Start collecting data for selected site(s).	Continue collecting data, use of data as relevant to monitoring and reporting (fully functional weather station)
Surface water hydrology and surface water modelling	Incorporate available information into the baseline information.	Start collecting regional historic data for hazard and environmental analysis; Start collecting data for selected site(s).	Continue collecting data, use of data as relevant to monitoring and reporting
Groundwater hydrology and groundwater modelling	Incorporate available information into the baseline information.	Start collecting regional historic data for environmental analysis; Start collecting data for selected site(s).	Continue collecting data, use of data as relevant to monitoring and reporting
Fauna vertebrates and invertebrates, fresh water ecology, wetlands ecology, botany	Initial assessment, could be desktop with limited site assessment/ground truthing	Detailed on-site assessment of impacts due to construction and operation, developing a baseline	Monitoring and reporting of impact of construction and operation against baseline; Implementation of offset if required.
Marine biology	Initial assessment	Detailed assessment of impact, developing a baseline	Monitoring during construction and in operation
Geohydrology	Initial assessment	Detailed assessment and monitoring over several seasons or years	Monitoring and reporting
Coastal/river/lake management (as appropriate)	Initial assessment	Detailed assessment of impact of construction and operation	Monitoring and reporting during construction and in operation
Demographics and sociology	Initial assessment	Detailed assessment	

TABLE I–1. TYPICAL AREAS OF EXPERTISE AND ACTIVITIES REQUIRED FOR ENVIRONMENTAL PROTECTION IN EACH PHASE (cont.)

Area of expertise	Phase 1	Phase 2	Phase 3
Archaeology, palaeontology	Initial assessment	Assess impact on known cultural sites and determine nature and likelihood of possible finds during construction, and develop a plan for dealing with findings.	Implement plan if necessary
Socioeconomic assessment	Initial assessment of demographics, stakeholder engagement	Detailed assessment for NPP project	Ongoing engagement and implementation of recommendations
Land use	Initial assessment	Detailed assessment	Implementation of recommendations
Environmental justice	Initial assessment	Detailed assessment	Implementation of recommendations
Traffic engineering	Initial assessment	Determine any new infrastructure needed; Assessment of impacts during construction and operation.	Development of infrastructure before and during construction
Visual impact		Detailed assessment	Implementation of recommendations
Noise impact		Detailed assessment	Monitoring and implementation according to legal requirements
Electromagnetic interference		Detailed assessment	Refine assessment and implementation of recommendations
Radiological dose/risk assessment	Initial assessment	Detailed assessment	Refine assessment
Radioactive waste	Initial assessment	Detailed assessment	Refine assessment
Emergency planning	Consider feasibility of emergency planning at candidate sites	Determine if emergency plan could be feasible at selected site	Prepare emergency plan
Legal	Gap analysis for national legislation	Prepare and implement necessary new legal instruments; Compile list of all existing relevant national legislation, international obligations and practice, lender requirements.	Prepare/review documentation and conduct/ enforce activities in line with applicable legislation and requirements.

TABLE I–1. TYPICAL AREAS OF EXPERTISE AND ACTIVITIES REQUIRED FOR ENVIRONMENTAL PROTECTION IN EACH PHASE (cont.)

Area of expertise	Phase 1	Phase 2	Phase 3
Grid	Initial assessment	Determine necessary new grid infrastructure, assess options (e.g. overground or underground and alternative routes) and impact (possibly separate EIA in parallel to site EIA)	Implement recommendations and monitor impact during construction and operation

I–3. TIMING OF INVOLVING SPECIALISTS

Some of the experts in these areas are involved throughout the EIA process. While it is recommended to involve specialists early in the process, the level of their input would vary depending on the phase of the project and the specific environmental aspects triggered at the selected site and through the infrastructure development on that site.

A number of issues need to be considered, including the following:

— The EIA report and any other submissions made will need experts, some to perform the assessment on behalf of the project proponent and others to review them on behalf of the regulatory bodies and decision makers. These experts are expected to be independent of each other and a system for managing conflicts of interest would be developed by the applicant and regulatory bodies. In some countries, the applicant / project proponent is required by the national regulations to appoint an independent environmental practitioner who would also assist with addressing any conflict of interest. The regulatory bodies would also be independent of the project developers if this is a governmental body. This may be especially difficult if there is a limited pool of experts available.
— If there is a shortage of expertise or in the number of experts available in some areas, support may be required from external or international organizations, consultants, etc. If this is the case, for either the developer or the regulatory body, an 'intelligent customer' capability will need to be developed so that the developer or regulatory body becomes knowledgeable enough to understand the basis of the analyses and conclusions drawn therefrom.
— The overall assessment needs to be balanced with the level of detail being only that needed (i.e. commensurate with the expected environmental impact or risk of impact). This will need careful management by the project developers to control expenses and avoid unnecessary work performed by experts in a particular field.

Some content that needs the input of specialists are necessary for both the environmental and radiological safety assessment. In such cases, careful management will be required to ensure that each area of assessment produces the necessary information for its own purposes and any other area of assessment that it supports. Specialties that support both the safety assessments and environmental protection include meteorology, hydrology, geology, geotechnology and economics. For each of these specialties, sometimes the same specialist or team of specialists support both the safety and environmental teams. At other times, different specialists are assigned to each team.

Meteorologists assist with determining the safety aspects of the NPPs in regard to severe weather such as storms, tornadoes, snow, dust and sand storms, hail, lighting and freezing precipitation. For EIAs, they are responsible for analysing the air quality related issues during construction and operation of the

NPP and for conducting the environmental transport analyses for radionuclides released to the air during normal operation and accident conditions.

Similar to meteorologists, hydrologists are responsible for analysing the effects of severe surface water related phenomena, such as flooding and tsunamis, on the safety of the NPP, and also for conducting the environmental transport of radionuclides released to environment in surface water bodies and in groundwater.

Specialists in the fields of geology and geotechnology analyse the stability of structures with systems that are important for nuclear safety. On the environmental side, they assist in setting up conceptual models for understanding groundwater regimes at and around the site for analysing the groundwater recharge and discharge, and for environmental transport of radionuclides in groundwater.

Economics will need to be considered from the outset in Phase 1, when deciding whether nuclear power should be part of the energy mix. In Phase 2 and Phase 3, economics will have a role in selecting the reactor technology and/or design or operational options. For example, options that reduce the environmental impact may cost significantly more; cost–benefit analysis or similar studies may be needed to arrive at an appropriate compromise and provide a justification for any choices made.

Annex II

TYPES OF UNCERTAINTIES WHICH CAN INTRODUCE PROJECT RISK

II–1. INTRODUCTION

Drawing conclusions about the impact an NPP project will have on the environment involves collection and analysis of environmental data as well as a thorough understanding of the reactor technology, particularly its potential releases to the environment. Additionally, it is expected that the interactions between the technology and the environment are known and fully understood. Since the life cycle of the plant spans several decades, estimates are made of the future environmental conditions, technology and interactions. All of these areas contain uncertainties. The methodology to address uncertainties may be described in a country's national regulations. The types of uncertainties which can introduce project risk are described below.

II–2. UNCERTAINTIES RELATED TO THE ENVIRONMENT

Data collection may not completely describe or may fail to describe all of the current environmental conditions, particularly with respect to subsurface conditions, biota life cycle descriptions (which are necessarily based on incomplete sampling schemes) or identifying sensitive species or habitats. Extrapolations to the future may not adequately consider environmental changes, due either to known issues or issues not currently known. For example, climate change could cause considerable changes to the environment that were not considered or accounted for during previous EIAs. An operating plant may need to revisit, at any time, the environmental information captured in EIAs in the light of newly identified issues.

To address uncertainties in the environmental data, assumptions could include:

— A bounding value for the dilution of any released radioactivity at a given point by atmospheric or aquatic transport such that any impacts calculated will be greater than the true levels (thus, the assessment will be a bounded value). Simple dispersion models or approaches could also be used, which tend to overestimate or use conservative values for unknowns; Safety Reports Series No. 19, Generic Models for Use in Assessing the Impact of Discharges of Radioactive Substances to the Environment [II–1], discusses the use of models of varying complexity that could be used in different stages of the process.

— A bounding value for the response of the environment to the released pollutant, such as the ability of the water body to absorb the thermal load.

— Consideration of the use of applicable international values for the habits of the local population (e.g. food consumption) where national data are not available. This can be supported by ongoing evaluation of local conditions, in some cases modelled and predicted.

II–3. UNCERTAINTIES RELATED TO THE REACTOR TECHNOLOGY

The reactor technology may not be identified when the EIA is initially prepared. Additionally, a project is usually bid on the basis of a reference plant which is operating elsewhere. In the process of designing the plant for the in-country site, changes to the reference plant design may be required, either to accommodate the site conditions or as a requirement of the regulatory body. Additionally, the technology design may be modified over the life cycle of the plant. All these changes may impact the analyses and results of the EIA and, therefore, the EIA would need to be re-evaluated considering the new information.

To address uncertainties in the technology, the PPE approach [II–2] may be considered, in which a value is selected that bounds those in the different reactor technologies that may be deployed for each plant parameter that will need to be considered in the EIA. The PPE includes the important physical and chemical parameters that may affect the environment (e.g. water requirements, land use and emissions) for the considered plants and identifies which technology has the value for a particular parameter with the highest potential impact value or range of values. For positive impacts (such as job creation or other positive socioeconomic impacts), the PPE parameter would be less than the project expected value (i.e. not as great a positive impact). The bounding parameters which are included in the PPE are then used for environmental analysis in the EIA process. When the final design is known, a comparison is made between the actual value for each aspect and the bounding value initially identified. If the ranges of actual values for the parameter are lower than or equal to (or greater than, for positive impacts) values on which the environmental analysis is based, then further environmental assessment is not required. Otherwise, a new environmental assessment will be required.

II–4. UNCERTAINTIES RELATED TO INTERPRETATIONS OF DATA AND CALCULATIONS OF IMPACT

Uncertainties in the input data and information will lead to uncertainties in the calculation of environmental impact. The level of uncertainty in the impact appropriate at any decision making point will depend on the irreversibility of the decision being made at that time and the level of risk — and where the risk lies — of the project ultimately being deemed unacceptable. For example, proceeding to the next stage of a desktop study can accommodate a high level of uncertainty, as only the developer's time and resources are at risk, whereas proceeding to site construction activities will impact the environment and expend significant resources. Consideration of variability and uncertainty in radiological EIAs is discussed in IAEA Safety Standards Series No. GSG-10, Prospective Radiological Environmental Impact Assessment for Facilities and Activities [II–3]. Modelling may be required for extrapolating the data set to the future or to address larger biota populations than those sampled, and the uncertainty known for that model would be reported.

II–5. UNCERTAINTIES RELATED TO REGULATORY CHANGES

Regulations governing the project may be subject to change at any time, such as in response to national considerations, new safety requirements or public pressure. To minimize the uncertainty associated with regulatory changes, the proponent of the project should engage in continuous meaningful communications with regulatory bodies and stakeholders.

Inevitably, uncertainties will remain even in the final assessment. These sources of uncertainty, for example, incomplete data or a changed design, should be identified, and their impact on the reliability and accuracy of the assessment should be evaluated in the EIA report. These impacts should also be monitored during the lifetime of the project to confirm the impacts are as expected.

REFERENCES TO ANNEX II

[II–1] INTERNATIONAL ATOMIC ENERGY AGENCY, Generic Models for Use in Assessing the Impact of Discharges of Radioactive Substances to the Environment, Safety Reports Series No. 19, IAEA, Vienna (2001).

[II–2] INTERNATIONAL ATOMIC ENERGY AGENCY, Managing Siting Activities for Nuclear Power Plants, IAEA Nuclear Energy Series No. NG-T-3.7 (Rev. 1), IAEA, Vienna (2022).

[II–3] INTERNATIONAL ATOMIC ENERGY AGENCY, Prospective Radiological Environmental Impact Assessment for Facilities and Activities, IAEA Safety Standards Series No. GSG-10, IAEA, Vienna (2018).

ABBREVIATIONS

EIA	environmental impact assessment
EMP	environmental management plan
EPR	emergency preparedness and response
ESR	environmental scoping report
NEPIO	nuclear energy programme implementing organization
NPP	nuclear power plant
PPE	plant parameter envelope
SEA	strategic environmental assessment
TSO	technical support organization

CONTRIBUTORS TO DRAFTING AND REVIEW

Altinyollar, A.	International Atomic Energy Agency
Avci, H.	Consultant, United States of America
Bastos, J.	International Atomic Energy Agency
Dubinsky, M.	Consultant, Israel
Geupel, S.	International Atomic Energy Agency
Haddad, J.	International Atomic Energy Agency
Harman, N.	Consultant, United Kingdom
Herbst, D.L.	Consultant, South Africa
Islam, K. R.	Rooppur nuclear power plant, Bangladesh
Kliaus, V.	Republican Scientific Practical Centre of Hygiene, Belarus
Kovachev, M.	International Atomic Energy Agency
Monken Fernandes, H.	International Atomic Energy Agency
Mursalova, G.	Ministry of Energy, Kazakhstan
Nunnini, L.	National Atomic Energy Commission, Argentina
Otabil, E.K.	Atomic Energy Commission, Ghana
Pigoulevski, M.	Belarusian NPP Republican Unitary Enterprise, Belarus
Sengezer, D.	Ministry of Energy and Natural Resources, Türkiye
Seydou, D.	Ministry of Mines, Niger
Shah, R.	International Atomic Energy Agency
Singh, J.	Nuclear Power Corporation, India
Stott, A.K.	International Atomic Energy Agency
Tabet, M.	National Office of Electricity and Drinking Water, Morocco
Telleria, D.	International Atomic Energy Agency
Walker, M.	International Atomic Energy Agency
Welsch, M.	International Atomic Energy Agency

Technical Meeting

Vienna, Austria: 5–8 October 2021

Consultants Meetings

Vienna, Austria: 23–25 February 2021, 24–26 August 2021

Structure of the IAEA Nuclear Energy Series*

Nuclear Energy Basic Principles
NE-BP

Nuclear Energy General Objectives
NG-O

1. Management Systems
NG-G-1.#
NG-T-1.#

2. Human Resources
NG-G-2.#
NG-T-2.#

3. Nuclear Infrastructure and Planning
NG-G-3.#
NG-T-3.#

4. Economics and Energy System Analysis
NG-G-4.#
NG-T-4.#

5. Stakeholder Involvement
NG-G-5.#
NG-T-5.#

6. Knowledge Management
NG-G-6.#
NG-T-6.#

Nuclear Reactor** Objectives
NR-O

1. Technology Development
NR-G-1.#
NR-T-1.#

2. Design, Construction and Commissioning of Nuclear Power Plants
NR-G-2.#
NR-T-2.#

3. Operation of Nuclear Power Plants
NR-G-3.#
NR-T-3.#

4. Non Electrical Applications
NR-G-4.#
NR-T-4.#

5. Research Reactors
NR-G-5.#
NR-T-5.#

Nuclear Fuel Cycle Objectives
NF-O

1. Exploration and Production of Raw Materials for Nuclear Energy
NF-G-1.#
NF-T-1.#

2. Fuel Engineering and Performance
NF-G-2.#
NF-T-2.#

3. Spent Fuel Management
NF-G-3.#
NF-T-3.#

4. Fuel Cycle Options
NF-G-4.#
NF-T-4.#

5. Nuclear Fuel Cycle Facilities
NF-G-5.#
NF-T-5.#

Radioactive Waste Management and Decommissioning Objectives
NW-O

1. Radioactive Waste Management
NW-G-1.#
NW-T-1.#

2. Decommissioning of Nuclear Facilities
NW-G-2.#
NW-T-2.#

3. Environmental Remediation
NW-G-3.#
NW-T-3.#

(*) as of 1 January 2020
(**) Formerly 'Nuclear Power' (NP)

Key
BP: Basic Principles
O: Objectives
G: Guides and Methodologies
T: Technical Reports
Nos 1–6: Topic designations
#: Guide or Report number

Examples
NG-G-3.1: Nuclear Energy General (**NG**), Guides and Methodologies (**G**), Nuclear Infrastructure and Planning (topic **3**), **#1**
NR-T-5.4: Nuclear Reactors (**NR**), Technical Report (**T**), Research Reactors (topic **5**), **#4**
NF-T-3.6: Nuclear Fuel (**NF**), Technical Report (**T**), Spent Fuel Management (topic **3**), **#6**
NW-G-1.1: Radioactive Waste Management and Decommissioning (**NW**), Guides and Methodologies (**G**), Radioactive Waste Management (topic **1**) **#1**